The Smithsonian Guide to

Seaside Plants of th

Gulf and Atlantic Coasts

from Louisiana to Massachusetts,
Exclusive of Lower Peninsular Florida

Wilbur H. Duncan
Professor Emeritus of Botany, University of Georgia

and

Marion B. Duncan

Smithsonian Institution Press, Washington, D.C., and London

The Wildcat Foundation has generously
provided support for this publication.

The book was designed by Carol Beehler
and edited by Eileen D'Araujo and Ruth
W. Spiegel.

Type was set by Graphic Composition,
Inc., Athens, Georgia.

The line drawings that accompany the text
are by Rose S. McAuliffe.

**Library of Congress Cataloging-in-
Publication Data**

Duncan, Wilbur Howard, 1910–
The Smithsonian guide to seaside plants of
the Gulf and Atlantic coasts from
Louisiana to Massachusetts, exclusive of
lower peninsular Florida.
Bibliography: p.
Includes index.
1. Coastal flora—Atlantic Coast (U.S.)—
Identification.   2. Coastal flora—Gulf
States—Identification.   I. Duncan,
Marion B.   II. title.
QK122.D86 1987   581.974   85-22095
ISBN 0-87474-386-9
ISBN 0-87474-387-7 (pbk.)

Cover: Eastern Prickly-pear [*Opuntia
humifusa*, 540].

Contents

For Mack, Lucia, and Douglas

Preface

Before the present century, man's primary means of global travel was by water and his first introduction to another land was an estuary, a cove, or a beach. Until the 20th century, the scenes viewed by early explorers to the North American continent remained relatively unchanged except for alterations dealt by nature itself. Damage by man was generally confined to areas of concentrated industry or scattered summer communities. Today an estimated 75 percent of the people of the United States live in states bordering oceans or the Great Lakes and population growth in these areas continues. Seaside recreation is available to greater numbers of people owing, in part, to their proximity to the coast as well as to convenient, rapid, inexpensive transportation and an affluence not previously enjoyed by so many. The resulting impact on the coastal environment gives cause for alarm and many areas are deteriorating rapidly under the increased pressure.

There are few new coasts for man to discover, yet each of us deserves the privilege of viewing and enjoying an unspoiled beach, a colony of dunegrass, a maritime woods. Our grandchildren deserve the same privilege. By presenting through photographs and descriptions some of the wildflowers, trees, and shrubs that currently grow along our eastern and southern seaboards, we hope that readers will be moved to aid in their preservation.

Acknowledgments

Information derived from interviews and field trips with laypeople as well as botanists and scientists in related professions contributed to this study. The many persons involved cannot be listed individually, but they are all warmly remembered. Paul Godfrey, University of Massachusetts; Lionel Eleuterius, Gulf Coast Research Laboratory; Cheryl McCaffrey, formerly with the Nature Conservancy; and Anne and Donald Bradburn were especially helpful. We are grateful to the artist, Rose S. McAuliffe, for providing the line drawings that accompany the text.

We are indebted to personnel of state parks and the National Park Service, at marine institutions and research stations, at universities, and to private real estate owners for their courteous assistance. We wish to express our thanks to the faculty and staff of the Botany Department, University of Georgia, for their extensive and continuous support. To the following friends we are indebted for the use of their photographs: Leo T. Barber, *Polygala incarnata*; Lionel E. Eleuterius, *Morus rubra*; Walter S. Judd, *Decodon verticillatus, Cicuta mexicana*; George Lewis, *Alternanthera*; Carol Ruckdeschel, *Tillandsia setacea, Epidendrum conopseum, Salix caroliniana*, and *Opuntia pusilla*; Donna J. Wear, *Aster divaricatus*.

Introduction

Authors' Note

This book is a treatment of seaside plants found on the Gulf and Atlantic coasts, from the Texas-Louisiana border to Plum Island, Massachusetts, except for lower peninsular Florida. The preparatory study for the book began in 1975 and included an extensive four-year survey of plants of the Gulf of Mexico coast from the Rio Grande at the Texas-Mexico border to Levy County, Florida, and the Atlantic coast from Flagler Beach, Florida, to Quoddy Point, Maine, and the Atlantic Provinces, Canada, exclusive of Labrador. Such an extensive geographic scope seemed justified to us, since these areas exhibited a surprising homogeneity of plant species and an actual pancoastal continuum of a remarkable number. The addition of lower peninsular Florida would have introduced a subtropical flora too different for compatibility; Labrador presented a coastline too formidable and inaccessible for practical consideration.

A treatment of that dimension, however, is physically limiting and, despite the overlap of species throughout the area, a need to decrease the geographic scope became apparent. Plum Island, Massachusetts, appeared to be a natural boundary for the upper Atlantic coast, as it represents the last of the major sandy barrier islands typical of the coastal plain. Shorelines of New England and the Atlantic Provinces as a whole are characterized by a series of drowned valleys separated by steep headlands that were formerly mountain ridges. Beaches, called shingle beaches, are narrow crescents composed of pebbles, cobbles, or boulders, or occasionally sand. Islands are craggy replicas of the mainland; coniferous forests extend down to the sea. Many plant species found on the Gulf and Atlantic coasts occur in this northern area, some extending the entire distance, but there is also a boreal flora not present southward.

Texas also contrasts to a significant degree with the rest of the study area. Although its coast is part of an extensive coastal plain framed by sandy barrier islands and many genera representative of the Gulf and Atlantic coasts are present, it is primarily a vast grassland. As in Florida, Texas also has many subtropical species growing in the extreme southern section that are not present or common in the other coastal states. We have, therefore, limited the Gulf states to those lying east of the Sabine

River, the boundary between Texas and Louisiana.

A book such as this has an additional limitation in that it cannot offer a complete floral accounting. Selection of species included has consequently been restricted to seed-bearing plants, excluding other members of the plant kingdom such as algae, fungi, mosses, liverworts, ferns, and fern allies. The 943 species chosen include showy specimens popularly regarded as "wildflowers," rare or distinctive species, common weeds, trees and shrubs, and easily overlooked grasses, sedges, and rushes. Many of the species occur in large areas of both Atlantic and Gulf coasts. A more complete treatment of the vegetation of the regions can be found in Correll and Johnston (1970) for Texas; Godfrey and Wooten (1981) for southeastern states; Radford et al. (1964) for the Carolinas; Fernald (1950), and also Gleason and Cronquist (1963) for northeastern states.

The definition of coast, as used in this study, is that area abutting the sea or strongly influenced by it. This includes island and mainland beaches, dunes, marshes, estuaries, and inlets as well as island interiors. With few exceptions, the entire area lies within the coastal plain; however, the coastal plain in its entirety is not involved.

To acquire a significant understanding of the North American eastern and southern seaboards, we traveled over 60,000 miles (96,000 km) within the original range of Texas to Newfoundland. We gathered maps of each county in that range and followed all dirt roads and paths to the beach along the entire range to be sure of complete coverage. Seasonal visits were essential and many species were, thus, observed during both flowering and fruiting stages. Repeated trips to selected sites lent a feeling of continuity to the study, and some places such as Cape Hatteras National Seashore, North Carolina; Jekyll Island State Park, Georgia; and the Mobile area of Alabama became familiar stops.

Plant treatments in this book derive primarily from our study of specimens and of published materials in scientific journals and recent floras. Botanists have been helpful in supplying us with results of their ongoing research, updating nomenclature, and adding to plant distributions. Herbarium and library research contributed documentation for subjects such as plant introductions, migrations, and decimation. Interviews with field biologists, park personnel, professional and amateur botanists, and other interested citizens aided us in locating specific species, directed us to interesting habitats, and provided us with valuable insights into the vegetational history of certain regions.

Our study has required attention to several disciplines other than botany and photography. The roles of geology, climatology, and other fields of natural history often explain the composition of plant communities and even relate them to other areas. Opening and closing of inlets, hurricanes, floods, fires, erosion, drought, and countless lesser natural phenomena can leave a record of vegetational alteration.

Cultural and Economic Notes

Current industrial and commercial practices are tending to interrupt the natural functioning of coasts. These include construction of jetties and groins, dredging of channels, draining of marshes, bulldozing of dunes and forests, and, of course, the building of communities and industrial plants.

Some of the earliest manifestations of interference by man with natural processes were caused by the grazing, grubbing, or browsing of domestic stock. The problem continues today, often due to feral animals. Although early man usually took only what he needed for survival, this was not necessarily due to environmental awareness. He was limited in what he could accomplish alone or with little assistance and his tools were primitive. Food and medicine, a means of transportation, shelter from the elements, and protection from animals and other men were the primary needs supplied by the environment, and the first Europeans who followed demanded little more. Shelters and fences reflected what was available in wood, thatch, and stone. Small boats were built from straight, fine-grained gums, hickories, and oaks. (The larger oaks were depleted later as ship-building flourished during the late 18th and early 19th centuries.) Clearing of forests for farms or pastures, for roadways, and for use as fuel and building material increased with the rise in population.

Grasses served early settlers in numerous ways. *Spartina patens*, commonly called marshhay or salthay, is a plentiful coastal grass that occurs from Texas to Newfoundland. It served in pioneer days, and continues to do so to a limited extent, as a natural pasture for stock and, until the 20th century, was extensively cut for hay. *Distichlis spicata*, known from Texas to Nova Scotia as saltgrass, was another prime constituent of marshhay. A second *Spartina* that occurs from Texas to Newfoundland, *S. alterniflora* (smooth cordgrass), was used as thatch for roofs. Grasses have also served needs other than as thatch or fodder. On several southern coastal islands, there are people who weave baskets of exquisite design and technique from a grass common to their dune areas and the margins of marshes, *Muhlenbergia filipes* (pink muhly).

Wax-myrtle (*Myrica cerifera*) is a shrub that occurs from Texas to New Jersey, often abundantly. In pioneer days, wax boiled from the berries of this plant was used in making candles, scenting soap, and as an air deodorant in colonial homes. In the far south, Spanish-moss (*Tillandsia usneoides*) has been used for upholstery and as a packing material.

Early coastal settlers learned from Indians, and by trial and error, plants that were efficacious as medicines and palatable or adequate as foods. Leaves of *Ilex vomitoria*, the yaupon holly that occurs from Texas to Virginia, were brewed into a black tea and used ceremonially by Indians. The tea in a weaker form was used for refreshment. Bark of sassafras roots, *Sassafras albidum*, gathered in the spring, was also brewed as a tea and drunk as a tonic as well as for refreshment. Unfortunately, use of plants as food or for medicinal purposes often requires total plant destruction and could lead to species eradication.

Plant Adaptations

Most coastal vegetation is subjected either regularly or occasionally to salt-water flooding or spray. In addition, a salt-water environment is created for many populations situated back from the shore by ocean winds that bear salt-laden water droplets or sand mixed with salt particles. In extreme instances of severe storms or especially high tides much vegetation is killed, but there are some species that survive. Since fresh water is a requirement of all plants performing the process of photosynthesis, the question arises as to how plants can function in a physiological desert of saline or brackish water.

Plants that can grow and reproduce under these stressful circumstances are called halophytes and are generally defined as those with a tolerance of 0.5% NaCl in soil water. They are unique in that they have adaptations, often highly specialized, for deriving their freshwater needs from salt water and for conserving the fresh water. One such adaptation is the presence of salt glands, organs found in stems and leaves that secrete the excess of salts dissolved in soil water. Some well-known salt community species having these glands include smooth cordgrass, salt-hay, saltgrass, and species of sea lavender (*Limonium*).

Succulence due to increase in size or abundance of certain internal cells provides additional storage space for water, as can be observed in leaves of saltwort (*Batis maritima*) and seashore-elder (*Iva imbricata*) and the stems of glasswort (*Salicornia*) species. Water loss through exposed leaf and stem surfaces is inhibited in some species by a heavy cutin layer coating these surfaces and retarding evaporation. This accounts for the waxy appearance of sea purslane (*Sesuvium*), railroad-vine (*Ipomoea brasiliensis*) and a number of other plants found along the coast.

Relatively recent research is revealing details of improved metabolic efficiency in a specialized class of plants in which water loss by photorespiration is drastically reduced. Other modifications include reduction in size or number of leaves; presence of special structures such as hairs or scales that protect stomates; recessing of stomates into leaf surfaces; opening of stomates during cooler night hours and closing during the day; reduction in number of stomates; and curling of leaf blades during hot daytime hours, thereby reducing total evaporation surface. Some plants regularly found in highly saline environments often exhibit more than one of the modifications contributing to salt tolerance. *Salicornia* species, typical of the salt community, have fleshy stems with a waxy cuticle and leaves severely reduced in size and number.

It must not be assumed that all plants exhibiting the characteristics just enumerated are necessarily halophytes or that they occur only along coasts. Many of these adaptations apply to other xerophytes, whether salt-related or not; and salt deserts can and do occur far inland. It should also be pointed out that many halophytes are adaptable and often inhabit nonsaline situations as well. An example is prickly-pear (*Opuntia humifusa*), a species that thrives along southern coasts behind primary dunes and equally well in rocky areas of the piedmont.

Adaptations other than to salt are often necessary for plants growing along exposed shores. The physical force of wind can be a conditioning factor, as winds are steady most hours of the day and can reach gale force during certain seasons. Camphorweed (*Heterotheca subaxillaris*) and beach plum (*Prunus maritima*) are examples of species that maintain a diminutive life form when growing on exposed coasts and a larger, fuller form in more favorable areas. Railroad-vine can trail for several meters along a berm or foredune, but does not expose an upright profile. Windshearing or flagging, a common feature of maritime live-oak forests in the south and of seaside woody vegetation worldwide, is attributable to wind action. Salt spray borne by the wind kills tissue on the windward side of the plant, causing a flag-shaped appearance as the leeward side continues to develop.

Coastal Topography

To better understand and appreciate the coastal vegetation, a more intimate acquaintance with the coast itself is helpful. Coasts are physical features common to all continents and islands of the world and their presence is generally accepted by the average individual with little thought or concern for their subtleties or uniqueness.

Shoreline is the interface between land and water, and coast is that zone landward of this boundary that falls under the influence of the sea, either directly or by other means such as salt spray or storm surge. Drift areas, dunefields, terraces, and seacliffs are included within the overall embrace of the term "coastal," as are estuaries and saline and brackish marshes and lagoons. Nor can the term exclude such nonsaline habitats as swales or interdunal slacks, forests, and freshwater ponds and marshes, all common features of the coast.

Changes over a long period of time that alter the appearance or position of seacoasts are closely associated with rising or falling sea levels and emerging or submerging land masses. Sea and land changes, in turn, reflect tectonic activity or climatic changes. There are several major processes affecting coastal configuration that are currently observable, often measurable, usually interrelated. Edaphic factors include such phenomena as sediment deposition, sand relocation, erosion, and accretion, and these in turn are related to hydrodynamic forces such as tides, currents, waves, and storms. Climate can be an influence for both of the above in the role of storms, temperature extremes, and weathering. Plants, man, and other animals introduce the biotic element affecting coastal configurations, with man the most manipulative of the three. Overall, there are broad similarities between all marine coasts, depending upon the nature and degree of interaction of formative and altering forces.

The Atlantic plain is a physiographic division that comprises the eastern and southern coasts of North America. This is a gentle slope to seaward that is divided into continental shelf (the portion lying submerged below the ocean) and coastal plain (the dry-land portion). From the Hudson River estuary to Texas, width of the continental shelf can vary in places from five miles (8 km) to around 100 miles (160 km). The coastal plain, which is more pronounced, ranges up to 300 miles (485 km) wide in portions of Texas. Much of the coastal plain includes low barrier islands that follow the continental trend in an almost continuous chain, separated from the mainland by shallow bays, marshes, or lagoons.

In the North Atlantic region from Long Island, New York, northeastward through New England and the Atlantic Provinces of Canada, there is no overall coastal plain. The Atlantic plain lies almost entirely underwater as continental shelf and can measure up to 300 miles (485 km) wide. This area is part of the region once covered by successive ice sheets, the most recent of which was the Wisconsin Glaciation during the Pleistocene Epoch. At the height of the Wisconsin Glaciation some 25,000 years ago, the mean sea level stood about 425 feet (130 m) below its present level and a broad coastal plain extended to the Grand Banks of Newfoundland. Glacial material clogged mountain valleys or was carried out to sea. With the warming trend starting about 12,000 years ago, glaciers melted and seas rose, drowning the coasts and flooding valleys for great distances inland. Morainal debris, transported by glaciers and deposited in masses, became islands, such as Long Island, New York, and Nantucket Island, Massachusetts, or land forms such as Cape Cod, Massachusetts. Islands and peninsulas were created from peaks and ridges by rising waters. Eventually, sea rise relative to the land began to decrease and has since leveled off to about 12 inches (30 cm) per century. Thousands of islands remain, however, rocky remnants of the mainland from which they were carved.

Barrier Islands

All coastal islands serve the common function of buffer to the mainland. They help dissipate the force of storm surges and absorb much of the energy of oceanic tides and currents, preventing or reducing the impact of floods and winds. Not only the mainland is afforded protection; harbors and estuaries in the lee of these natural barricades escape much of the extreme forces directed landward from the oceans.

Barrier islands of the coastal plain contrast as a whole with those of the glaciated areas, not only in origin and relief, but in their interactions with the sea. It is possible to formulate a generalized description of them despite the individuality they can exhibit, not only regionally but within a related chain. For the most part they are relatively young in geologic age (mostly less than 10,000 years old), they are elongate and narrow, and they parallel the mainland coast. Sands of the beaches and dunes are unconsolidated and constantly shifting. Maritime dominance has the salutary effect of creating a generally equable climate with persistent breezes and limited temperature extremes.

Salt marshes characteristically line the back or lagoonal side of a barrier island. In the Gulf of Mexico from Florida to the Mississippi Delta, however, open water separates most islands from the mainland, sometimes for a distance of up to 10 miles (16 km) or more. Here marshes occur prominently along the shores of the mainland, and sparsely on the islands. These islands lie in an active hurricane zone and are periodically lashed, overwashed, and frequently severed by these powerful storms.

Barrier islands cannot function with total independence; no single member can be isolated as a discrete entity. Each is part of an overall system, interrelated, involving water and wind currents, sediment transport, and sand budgets. Storm surges that carve an inlet through an island release that displaced sand for reworking and deposition elsewhere, usually in lagoonal marshes or as new accretion on another island downdrift. In a similar manner, overwashes and blowouts recycle sands, availing them once again for deposition elsewhere within the system. As a natural geomorphic construction, barrier islands are dynamic and constantly subject to alteration by natural forces. It is, therefore, not difficult to understand how one local manipulation by man can initiate a chain of reactions that ultimately may alter an infinitely larger area.

Wetlands

Wetlands that are regularly inundated and drained by tidal action are a characteristic of most coastlines and are especially associated with estuaries and lagoons. In intertidal zones throughout the world, wetlands have broad similarities. They develop in muds and sediments on lands either recently submerged by rising sea levels, as along the New Jersey coast, or lands that have recently emerged where coastlines are rising, as in western Newfoundland. The soil is water-saturated, high in salts, and low in oxygen. These wetlands support large areas of vegetation of relatively low diversity, possibly due to the fact that few plants can survive regular periods of tidal submergence or saline soils with low oxygen, thereby reducing competition.

Composition of the wetland flora varies according to soil and climate.

In latitudes near the equator, both north and south, several species of mangrove constitute the principal growth, with few, if any, herbaceous species present. On the Atlantic coast in both hemispheres where temperate or colder climates prevail, mangrove swamps are replaced by salt marshes.

North Atlantic salt marshes generally grow on a peaty substrate along tidal inlets and behind offshore islands and spits. Here smooth cordgrass dominates the low marsh, succeeded in the higher marshes by salthay, prairie cordgrass (*Spartina pectinata*), and common reed (*Phragmites australis*). Marshy areas that receive a preponderance of fresh water from the interior support populations of common cattail (*Typha latifolia*), tule (*T. domingensis*), and slender blue-flag (*Iris prismatica*).

Salt marshes of the mid-Atlantic coast and temperate latitudes of the South Atlantic and the Gulf of Mexico generally resemble each other in composition. Smooth cordgrass and perennial glasswort (*Sarcocornia perennis*), essentially alone, occupy the low marsh where tidal submergence is greatest. Black-grass (*Juncus gerardi*) or needle rush (*J. roemerianus*), salthay, and species of glasswort form dense stands landward of and adjacent to the first two. Species that grow in the high marsh where there is less flooding include coastal dropseed (*Sporobolus virginicus*), saltgrass, sea ox-eye (*Borrichia frutescens*), and species of spike-rush (*Eleocharis*), lilaeopsis (*Lilaeopsis*), glasswort, sea-lavender, fimbristylis (*Fimbristylis*), and bulrush (*Scirpus*). The highest level of brackish extent is frequently a shrub border of silverling (*Baccharis halimifolia*), false-willow (*B. angustifolia*), and marsh-elder (*Iva frutescens*).

Tidal waters drain away incompletely in some higher elevations that are reached infrequently by the tides and evaporation occurs on the site. Barren salt flats, or pans, are thus formed on which even the most salt-tolerant species are unable to survive. The first species to be encountered along the margins of flats as salinity decreases are saltwort, coastal dropseed, and species of glasswort, followed by species generally associated with the high marsh—cordgrass, sea ox-eye, aster (*Aster*), fimbristylis, and sea-lavender.

There are areas protected from the tides by sand ridges or dunes, or where tidal flushing occurs irregularly. Here brackish water can accumulate; the fresh water originating inland becomes mixed with salt water during spring tides and storms. These brackish wetlands support grasses, sedges, and rushes that are usually species of *Spartina*, *Zizaniopsis* (southern wildrice), *Fimbristylis*, *Scirpus*, or *Juncus*. A wide variety of forbs also flourishes in this environment—species of *Typha*, *Pluchea* (marsh-fleabane), *Sesbania* (sesbania), *Bidens* (beggar-ticks), *Oenothera* (evening-primrose), *Limonium*—together with pioneer woody plants, typically *Baccharis*, *Myrica*, *Juniperus* (red-cedar), *Cephalanthus* (buttonbush), *Acer* (maple), and *Salix* (willow).

Wetlands account for millions of acres of rich productive coastal property that is valuable to individuals, to governments, and to society as a whole. Studies have been made within the past two decades to determine marsh values in monetary terms in an effort to communicate with members of the business community in a common terminology. Gosselink, Odum, and Pope (1973), in a report on research conducted by themselves and colleagues on Georgia and Louisiana marshes, enumerated the beneficial functions performed by an average estuarine-lagoonal ecosystem. They used formulas employed in regular business practice and affixed a dollar value for each function. Their findings took into account that marshlands are a renewable resource whose income is a continually increasing stream extending into the future, as well as the often-overlooked fact that some of the real value accrues at some distance from the marsh itself. Aside from esthetic and other intangible factors, a convincing case was made for protecting the free service offered by nature. High among these services was the role of estuarine marshes in the assimilation of wastes in which nutrients are stored and recycled without appreciable effect on water quality. Also cited was the role of marshlands as a vital breeding site and nursery for shellfish and finfish, with the resulting contributions to commercial and sport fishing. Natural, undisturbed coastal wetlands are an integral part of man's total environment whose value is only now being slowly recognized.

Beaches

The image most often conjured in the public mind at the mention of the coast is a wide sandy beach and bordering dunes. Beaches occur where coastal currents and waves transport and deposit along the open shoreline materials that accumulate and widen into a strand, forming a fringe of land that is regularly exposed at low tide. It is a fluctuating belt that is easily cut and reduced during periods of storm and rough seas, then rebuilt during times of relative calm, as during summer months. In the instance of southern and some northern beaches, this transported and deposited material is sand and fine gravel eroded from some other coastal site or brought down by rivers or glaciers to the continental shelf. Not all beaches, however, are composed of sand; some are shingle, made up of rock fragments ranging in size from pebble to cobble to boulder. Such beaches are a common feature along shores in upper New England and the Atlantic Provinces and may form wherever sand sources are limited or wave energy is sufficient for the transport of large sediments. Finely crushed shells, called shell hash, may be found on any beach with a shell source at hand, but is more commonly a component of southern beaches.

The first line of sand-beach vegetation begins as a sparse and spotty assortment of hardy species whose seeds or vegetative parts have lodged in the accumulated detritus at high-tide mark. Sea-rocket (*Cakile*), reported to be the first vascular vegetation to appear on Iceland's newly formed volcanic island of Surtsey, is recognized globally as a pioneer genus in harsh environments. Sea-beach sandwort (*Honkenya peploides*), beach amaranth (*Amaranthus pumilus*), and species of Russian-thistle (*Salsola*) and sea-beach atriplex (*Atriplex*) are also frequently recognized at or near the drift line.

Lying behind, or landward of, a beach is the berm, a fairly level expanse of loose sand, varying in width, escaping all but the highest tides, and subject to frequent salt spray. It is the backshore and abuts the duneline or natural seawall at its border away from the beach. Vegetation here exhibits a greater variety than that found on the driftline and is more closely associated with dune plants. In this area incipient dunelets form as soil-binding dune grasses and forbs such as morning-glory (*Ipomoea*), croton (*Croton*), wormwood (*Artemisia*), sea-purslane, glasswort, and sea-shore-elder become established.

Dunes

Dunes are mounds of unconsolidated sand and are more familiarly associated with barrier beaches, where they lie close to the seaward shore. They are not restricted to this location, however, and can form on the lagoonal side or in the interior of islands, in the interior of a continent hundreds of miles from the sea, on the mainland shore, or wherever winds of sufficient energy have access to an adequate supply of sand.

Coastal dunes are formed by wind blowing across a beach, lifting sand from its surface, later losing this sand where plants or other impediments reduce the force of the wind, causing it to release its sediments. Uninterrupted, this process continues and dunes of considerable height can result.

The principal agents contributing to maintenance of the dune in its position are grasses that possess extensive root and rhizome systems and have the ability to produce new growth following sand burial. The most effective grasses are American beachgrass (*Ammophila breviligulata*) in the North Atlantic region and south as far as Virginia, and sea-oats (*Uniola paniculata*), seaside panicum (*Panicum amarum*), sandbur (*Cenchrus tribuloides*), and salthay in the southeast.

Dunes protect island interiors from primary exposure to oceanic forces, but continuous unbroken dunelines such as those constructed artificially in some areas fail to function harmoniously with the dynamics of barrier island systems. On a natural marine shoreline there are breaks in the dune wall where ocean surges can flow through and dissipate on the flats beyond. Where there is no such escape route, wave energy more readily erodes the beach, undercuts the dunes, and a natural breakthrough ultimately occurs.

Wind, the force that creates a dune, can also destroy it. Action becomes initiated at some spot where vegetation has been damaged or killed. This can be due to tracks or trails made by people, animals, or offroad vehicles or because of natural phenomena, weakening the dune to the extent that erosion ensues and blowouts ultimately result.

Where dunes remain relatively undisturbed and sand continues in adequate supply, vegetation increases to include forbs that further bind the soil. Pennywort (*Hydrocotyle*), camphorweed (*Heterotheca*), and seaside spurges (*Chamaesyce*) are early colonizers along most coasts together with members of the rose and heath families in the north and legumes in the south.

In the natural order of succession as stabilization of the dunes proceeds, shrubs and vines follow and then trees. Beach-heather (*Hudsonia*), and bearberry (*Arctostaphylos*) are chamaephytes (subshrubs) that colonize stable or protected dunes of the North Atlantic coasts in advance of larger shrubs. Species of myrtle (*Myrica*) and juniper invade over a wide area; poison-ivy (*Toxicodendron radicans*), species of grape (*Vitis*), greenbrier (*Smilax*), and other vines have general distribution; and redbay (*Persea*), holly (*Ilex*), buckthorn (*Bumelia*), and yucca (*Yucca*) are familiar genera on southern coasts. Forests, the ultimate cover of stabilized dunes of the Atlantic coasts and parts of the Gulf, show possibly the sharpest variation geographically of any of the coastal vegetation. Woods of the maritime strands of the southern Atlantic states consist conspicuously of evergreen oaks (*Quercus*) and often palms (*Sabal*); in states of the eastern Gulf region species of pine (*Pinus*) uniformly occur over large areas; where shell middens occur, magnolia (*Magnolia grandiflora*), live oak (*Quercus geminata*), and cabbage palm (*Sabal palmetto*) are frequently found. Deep, more protected forests throughout the range support selections of species from genera such as plum (*Prunus*), hickory (*Carya*), maple, oak, pine, and holly.

Flatlands

The area behind the dunefield is usually a flatland and is maintained as such by flooding from high seas or by overwash. Flatlands normally exhibit a herbaceous plant cover although shrubs will invade if the interval between floodings becomes lengthy. Along the Gulf coast, and particularly Texas, these flats support extensive grasslands with a wide range of species represented. Use of these grasslands as pasture for grazing stock is widespread. Meadows or savannas are more characteristic of Atlantic coasts with forbs added to grasses and sedges as a common component.

The water table in the flatlands is close to the surface and frequently rises above it, forming swales or slacks. These may be of sufficient depth to form pools or shallow ponds and may be fresh water if adequately protected. Freshwater marshes occur around the borders of these ponds or may completely fill swales of shallow depth. Plants living in these habitats are adapted to the climatic conditions characteristic of the coast, but may succumb when accidentally inundated by salt water during storms.

Freshwater marshes typically exhibit a flora composed prominently of *Fimbristylis, Cyperus, Scirpus, Eleocharis,* and other members of the large sedge family. Representative genera of other families commonly occurring with the sedges include *Juncus, Panicum* (panic grass), *Erianthus* (plumegrass), *Mikania* (climbing hempweed), *Typha, Boehmeria* (false-nettle), and *Gratiola* (gratiola). Where pools of a generally permanent nature exist, an aquatic flora is often introduced that is complex in nature and is not included here. Aquatic genera of pool margins in the south may include rushes, cattail, water-hemlock (*Cicuta*), water-hyssop (*Bacopa*), water-purslane (*Ludwigia*), arrow-head (*Sagittaria*), and pickerel-weed (*Pontederia*); their northern counterparts often include loosestrife (*Lythrum*), water-horehound (*Lycopus*), beggar-ticks, cattail, and knotweed (*Polygonum*). As shrubs and trees invade, swales become swampy, shading out many herbaceous species. Members of the heath family are early shrub invaders; other typical woody plants of this habitat include alder (*Alnus*), willow, and gum species.

Survey of Coasts by Regions

Subdivision of the coast into regions based upon modifying phenomena is commonly practiced by oceanographers, geologists, botanists, and coastal specialists in many fields for purposes of understanding physical features, island dynamics, vegetational differences and other aspects that lend individuality to an area. Regional differences result from varying reactions to regulative forces that mold the coasts; these forces are in turn influenced by geographic location. For example, the western shores of the Gulf of Mexico lie in the path of prevailing winds entering through the Florida Straits and exhibit a shoreline molded by strong wind and tidal action. Western Florida's Gulf border, by contrast, is a low-energy coast lying in a sheltered arc normally by-passed by major wind action. In a somewhat similar analogy, the Georgia bight or embayment abuts an 80-mile-wide (130-km) continental shelf, whereas Cape Hatteras has a continental shelf under 10 miles (16 km) in width. The effects of waves traversing a long fetch of shallow sea as opposed to a short one naturally vary. The divisions used in this study are presented in sequence, beginning with the northern coast of Massachusetts.

Massachusetts to the Hudson River Estuary

Massachusetts, Rhode Island, Connecticut, and New York are part of the larger glaciated region extending from Maine to the Hudson River that is characterized as drowned coastline. It is a coast of embayments, coves, prominent capes, and myriad islands and peninsulas. In Massachusetts,

Cape Cod Peninsula and several of the large islands are constructed primarily of glacial debris. Plum Island, however, and the Province Lands of Cape Cod as well as many spits, necks, and islands off Cape Cod are either Holocene or of Holocene construction built on underlying Pleistocene glacial deposits. They more nearly resemble the barrier beaches to the south than the shingle beaches northward and it is because of these similarities that they are included here.

Massachusetts and Rhode Island have much industry that is marine-oriented; both are urban in development with high-density populations, and the natural areas are used primarily for recreational purposes. Long Island Sound forms the marine boundary for Connecticut and the northern border of Long Island, New York, providing both with protection from direct exposure to the Atlantic Ocean. This portion of Connecticut has undergone extensive commercial, military, and residential development, utilizing most of the waterfront. Fifty percent of the marshland has been destroyed, with more already marked for dredging or filling. Fortunately, state parks such as Sherwood Island, Hammonasset, Harkness, and Bluff Point were established in time to preserve a few segments.

The northern coast of Long Island is somewhat similar to that of Connecticut in configuration, but has tended to attract residential development rather than heavy industry or maritime commerce. Holocene deposits along the Atlantic-facing south shore have produced a system of narrow, sandy spits running parallel to the shore almost continuously, with occasional inlets providing access to the bays. Development and heavy public use have altered much of the island property close to New York City, and population pressures are likely to extend this threat. Erosion is prevalent in some areas, but miles of low, broad dunelands remain along spits such as Fire Island where they reach imposing heights in some areas at the eastern end. Great South Bay, Peconic Bay, Gardiners Bay, and numerous other bodies of water provide marshes and similar habitats that contribute toward island stability.

Although the coasts of these four states are characterized by high population densities and extensive commercial development, there are spots to be found where such usage is prohibited, where damage from misuse is being controlled, and where the original natural functions and beauty are returning. One of the most outstanding examples is Plum Island on the upper coast of Massachusetts near the New Hampshire border. This island, like so many near urban areas, has suffered in the past from indiscriminate use, and the northern third remains in private development. However, most of the southern two-thirds was federally acquired in 1944 and converted to a permanently protected natural area known as the Parker River Wildlife Refuge. The Commonwealth of Massachusetts also owns a small park at the southern tip. Management practices in the Refuge have been geared to encourage continued use of the island by waterfowl, for which the island is well known, and to re-establish environments for native plant communities. A diverse flora is possible, as Plum Island exhibits a typical barrier profile of beaches, dunes, and wetlands. Rugose rose (*Rosa rugosa*), beach wormwood or dusty-miller (*Artemisia stelleriana*), and beach-pea (*Lathyrus japonicus*) are established in the foredune area, colonies of beach-heather dot the interdunes, and a shrub zone on the secondary dunes includes meadowsweet (*Spiraea latifolia*), and arrow-wood (*Viburnum dentatum*). On Plum Island, paper birch (*Betula papyrifera*), a northern species, approaches its southern limits along the coast.

Another spot of unique beauty is Napeague Beach in the vicinity of Montauk Point at the extreme eastern end of Long Island. High dunes rise behind the sandy to gravelly shelving beach, which is terraced to the extent that only the highest tides reach the dunes. Beachgrass and beach-pea are the principal vegetation on the primary dunes. Behind the first dunes beach-heather occurs in scattered low spots, often in pure stands. Here, also, large circular depressions in the sand are completely covered with spreading colonies of beach plum (*Prunus maritima*) in stunted condition, some only 6 inches (15 cm) tall and producing fruit. Beach-heather continues on back dunes, mixed with bearberry (*Arctostaphylos uva-ursi*), which in places forms a total ground cover. The low branches of pitch pine (*Pinus rigida*) become buried in sand, take root, and continue to spread, covering the major portions of a dunelet. A denser shrub zone is at some distance back from the beach where dune patterns are irregular and with no apparent orientation with the shoreline.

Parts of this area are under state protection as a park, and, since the press of public use is less severe here than it is nearer New York City, the prospects are good that these natural conditions will continue.

Hudson River Estuary to Chesapeake Bay

The Atlantic seaboard from Long Island to Virginia Beach is a low-lying plain, once the ocean floor, with a recent geologic history of renewed submergence. Most of the shoreline is bordered by a fairly continuous chain of low barrier islands that parallel the mainland, from which they are separated by lagoons and bays. The area derives much of its character from two major embayments: Delaware Bay and Chesapeake Bay. The interior shores of these bays are not treated here.

Delaware Bay forms a marine boundary for southern New Jersey and this, together with the long eastern Atlantic coastline and the Delaware River boundary to the west, separates New Jersey into a semi-peninsular state. The 125-mile (200-km) Atlantic shoreline begins its northeast boundary with a sandy hook curving upward into Lower Bay of New York Harbor followed by a chain of barrier islands that continues the length of the state. Most of the islands show signs of erosion and disturbance caused by urban development and extensive industrialization.

Island Beach State Park lies about 50 miles (80 km) north of Atlantic

City. This 10-mile-long (16-km) island is a remarkably preserved remnant of the beautiful barrier islands that once graced the New Jersey coast before urban encroachment altered them. Although not totally free of disturbance, this unusual island has been protected as a wildlife sanctuary and nature preserve since being purchased by the state in 1953, and before that was part of a private estate. Freshwater swales behind the foredunes, sandy pockets of beach-heather in the back dunes, shrub and tree zones in the stabilized dunes, and extensive wetlands can be found within the short half-mile (0.8-km) width extending from the ocean to the backbay. Many of the native species are rare on the island, and some are restricted to habitats that are unusual or infrequent.

Delmarva Peninsula, composed of the state of Delaware and portions of Maryland and Virginia, is a 200-mile-long (320-km) peninsula suspended between the Delaware and Chesapeake Bays and the Atlantic Ocean, separated from the mainland by a narrow neck to the north scarcely 10 miles (16 km) wide. Most of the Delaware coast lies close to metropolitan centers and has undergone general conversion into weekend and summer resort facilities. Maryland has experienced similar resort construction as far south as Ocean City, with few natural vestiges remaining. Below Ocean City, federal and state procurement of barrier islands from private ownership has checked development except for a few isolated spots. The upper 6 miles (10 km) at the north end of Assateague Island are completely blighted by crowded housing for summer use, but the remainder of this 33-mile-long (53-km) island is in a seminatural condition. Assateague State Park occupies the next 2 miles (3 km) below the summer colony and the remainder of the island is a National Seashore. Chincoteague National Wildlife Refuge adjoins it in Virginia.

The northern portion of Assateague Island National Seashore presents an interesting profile of broad flats spreading extensively between primary dunes on the east and a shrub zone along the bay front. These flats vary from fields of sandy dunelets to sterile salt pans. Big Fox Level is a wide, barren sand flat that is under water from rain most of the winter. Many swales are pink in July with marsh-pink (*Sabatia stellaris*) and swamp rose-mallow (*Hibiscus moscheutos*).

Lying off Virginia's eastern shore, stretching 50 or 60 miles (80–97 km) to the mouth of Chesapeake Bay, is a filigree of 18 sizeable barrier islands, unbridged and remarkably undefiled. Thirteen of these islands, called the Virginia Coast Reserve, are owned totally or in part by The Nature Conservancy, whose expressed intent is to preserve them in their natural state. Of the remaining five islands, four are protected by federal or state governments; only one is in private ownership.

Chesapeake Bay to Cape Romain

Coastal Virginia south of Chesapeake Bay is an urban complex developed around commercial and military interests that have absorbed the resort city of Virginia Beach. Below Virginia Beach, and in sharp contrast, Back Bay Wildlife Refuge extends to the North Carolina border. It is a narrow spit of low land washed on the east by the Atlantic and the west by Back Bay and frequently overwashed by the ocean during intensive storms. A continuation of the spit into North Carolina is called Currituck Banks.

From here, the Atlantic seaboard bears strongly southeastward, and the chain of barrier islands reaches its easternmost limit at Cape Hatteras. These islands are long, narrow, sandy, low-lying, and subject to frequent overwash. Off Cape Hatteras the continental shelf is about 5 miles (8 km) wide; the Gulf Stream leaves the course it has been following along the edge of the shelf and turns seaward. Here, too, Pamlico Sound becomes an inland sea with a width of about 30 miles (48 km).

Hatteras is the most northern of a series of capes, separated by embayments, extending to Cape Romain in South Carolina and including Capes Lookout and Fear in North Carolina. An extensive study of this cape-embayment pattern by Hoyt and Henry (1971) establishes some general facts applicable to all: the capes are barrier islands of Holocene origin (10,000 years or less); their locations are related to the discharge of major river systems either presently or historically; all have shoals extending to sea for several miles; and they probably represent a former, more extensive, system of capes that has undergone erosion and retreat.

In the upper half of North Carolina, the barrier chain is separated from the mainland by large brackish bodies of water with sandy bottoms and shallow depths except at locations of major river channels and inlets. For the remainder of the North Carolina coast and the upper third of the South Carolina coast, the island-mainland border narrows to lagoons, tidal creeks, and marshes.

As a whole, the barrier islands that extend from the Virginia border south for 140 miles (225 km) to Cape Lookout are called the Outer Banks. Currituck Banks is that portion of the Outer Banks that extends from the Virginia border 25 miles (40 km) south to Nags Head. The vagaries of storms and shifting sands alternately cut or close this long spit of land, causing it to remain in a relatively undeveloped state. Two large segments of the Outer Banks are designated National Seashore: Cape Hatteras National Seashore extends from Whalebone Junction to Ocracoke Inlet, and Cape Lookout National Seashore occupies the southern end of the Banks from Ocracoke Inlet to Beaufort Inlet.

Between Currituck Banks and Cape Hatteras National Seashore there are several small communities that are popular as fishing and resort centers and some, such as Kitty Hawk and Nags Head, have considerable historical significance. From Beaufort Inlet to the metropolitan complex of Myrtle Beach, South Carolina, extensive resort development has oc-

curred, ostensibly without plan or discipline, leveling dunes and maritime forests in the process.

Cape Hatteras represents a transitional zone for vegetation created
by factors related to its geographical position. It is here that northward-
flowing warm waters of the Gulf Stream meet cold waters of the coastal
currents from the north, strongly influencing the climate. The continental
shelf is narrow, with shoals extending eastward for some distance, affecting tide and wave action. This can be especially profound during periods
of storm. Seeds and viable vegetative parts carried along by prevailing
offshore currents have here a final opportunity for becoming established,
accounting for regional limits of certain plants and variations in species
associations. Nags Head Woods north of Cape Hatteras is primarily a
hickory-beech community; Buxton Woods near the Cape is a pine-oak-
hickory community. Highbush and lowbush huckleberry occur on Nags
Head; there are no heaths on Hatteras Island. Beach-heather has its
southern limits on Nags Head; no naturally occurring palm species extends farther north than Hatteras.

Buxton Woods has the greatest species diversity of the Outer Banks,
according to Clay Gifford, a veteran interpretive naturalist with the National Park Service. It is a 9-square-mile (23-km²) area established on old
stabilized dunes with occasional interdunal freshwater ponds. A rich herbaceous flora includes several species of orchids and the understory
shrub growth includes *Sabal minor*, the palmetto.

Cape Romain to St. Johns River Estuary

The lower two-thirds of the South Carolina coast and a portion of northeast Florida are treated as a unit with Georgia's coast, which they greatly
resemble in physical appearance as well as geologic history. Offshore islands, locally called sea islands, frame the coastline. They contrast significantly in appearance with the long, narrow overwash formations directly
to the north and south in that they are considerably broader in relation to
their length. They are comparable to islands of New England in origin in
that they are drowned-mainland products of the late Pleistocene Epoch;
however, they contrast in that they have developed along an alluvial
rather than a rocky shoreline, and they have extensive Holocene deposits
on the ocean side. A few of the smaller islands are totally of Holocene
age. Well-developed dune fields front the ocean, flats and swales occur
between the dunes, and mixed forests dominated by oak or pine occupy
the interiors of all but the smallest islands.

Navigable waterways separate most of the island chain from the
mainland. Vast marshlands, often 4 to 6 miles wide (6–10 km) and cut
intermittently by meandering creeks, border the islands or the mainland,
or both, along these waterways. Southerly littoral currents are actively
eroding the northern ends of some of the islands, with a resulting

buildup of beaches and spits along the accreting south ends. The continental shelf is broad, up to 80 miles (130 km) at its broadest point off Brunswick, Georgia, and shallow with a sandy gently sloping bottom. Wave energy is relatively low level with only slight seasonal changes. The tidal range is wide.

Much of South Carolina's lower coast is influenced by the harbor at Charleston. Regular dredging of the harbor is necessary due to silting from the Ashley and Cooper Rivers that form the estuary and shoaling from sediments deposited by southwesterly currents. This has resulted in extensive eroding or accreting of nearby islands.

Cape Romain National Wildlife Refuge, and a few state parks, administer portions of South Carolina's lower coast, but much of it is privately owned. Around Charleston, residential, industrial, and military growth have caused the metropolitan area to spill beyond city limits to nearby inlands.

Georgia's "Golden Isles," like those of adjacent South Carolina and Florida, have felt the influence of human occupancy for the past 4,000 years. Plantations were established on several of the larger islands during the 18th and 19th centuries and prospered in the export of agricultural products until the Civil War.

Fifteen major islands flank the Georgia coast. Four of these, privately owned St. Simons, Sea Island, and Tybee, and state-owned Jekyll, have bridges to the mainland. They, thus, serve as bedroom communities for nearby cities and as recreational resorts. Of the four additional privately owned islands, Little Tybee is developed, Little St. Simons is being maintained in a natural state, and St. Catherines and Little Cumberland belong to nonprofit organizations. The remaining seven are in state or federal ownership, including Cumberland, which is a National Seashore.

The pressures for development versus forces for conservation and wilderness preservation have found admirable compromise in two islands of this general region, Kiawah Island, South Carolina, and Little Cumberland Island, Georgia. Kiawah is bridged to the mainland whereas Little Cumberland is not; they operate under dissimilar administrations, yet both have achieved similar results in appeasing the needs and demands of these two disparate philosophies. The owners of Kiawah based their projected plans for island development on a privately commissioned environmental impact study conducted by professional scientists. In keeping with this plan they have designated one-third of the island for residential and support use and one-third for recreation; the remaining third is to remain wild. Construction has been restricted to areas behind the foredunes and foot traffic is limited to boardwalks constructed over the dunes, around ponds, and through swamps. If development continues faithful to plans, Kiawah will illustrate the sane handling of the real problem of recreational need versus wildlife preservation.

Little Cumberland Island has attracted interest over a span of years because of its excellent beach within sight of an urban area. During its early stages of settlement, property owners formed a corporation and outlined stringent guidelines for land use. As a result, the island remains remarkably unspoiled despite the number of houses that have been allowed.

In Florida, Amelia Island is a large sea island north of Jacksonville that extends to the St. Marys River estuary. Originally it resembled the mainland in its dense, hammocky vegetative cover of magnolia (*Magnolia grandiflora*), redbay (*Persea borbonia*), and cabbage palm. Recently, however, extensive resort and retirement community development has occurred, depleting much of the native vegetation. Fort Clinch State Park at the northern tip of the island remains in a moderately natural state and is well situated for fishing as well as beach activities.

St. Johns River Estuary to Flagler Beach

The Gulf Stream, or Florida Current, flows northward a short distance off Florida's Atlantic coast, contributing to the prevalent mild Caribbean climate. Flowing in the opposite direction and nearer the shore, a strong longshore current from the north washes the coast, bringing with it sediments of quartz sand from the Appalachian highlands. Extending from the St. Johns estuary to Flagler Beach is a continuous line of narrow barrier islands and bars with wide beaches of quartz sand and shell hash fronting the ocean. Dunes of almost pure quartz sand rise the entire length except where artificially removed or damaged. Salt marshes border the narrow, shallow lagoons on the west side of the islands and along the mainland. Near Fort Matanzas there is a black mangrove swamp where *Avicennia germinans* reaches its northern limit along the eastern coast.

The combination of mild climate and wide, gently sloping beaches of fine white sand have made Florida's east coast an ocean playground, with tourism a major resource. Before barrier island dynamics was fully understood, almost continuous strip development had occurred with damage of lasting proportions resulting from dune destruction and eradication of natural vegetation. Fortunately, Florida recognizes the dimension of its coastal problem with its inherent threat to the tourist trade and the state economy as a whole, and legislated management plans are in progress.

Northwest Florida from Cedar Key to Alligator Point

The Gulf of Mexico forms a marine boundary for approximately half of the southern United States and most of eastern Mexico. Its almost continuous shoreline embraces the states of Florida, Alabama, Mississippi, Louisiana, and Texas in the United States, and five states in northeastern Mexico. A broad continental shelf, extending from a similarly broad coastal plain, forms an uninterrupted terrace except at the Yucatan Channel and along the coast of Veracruz, where recent geologic activity has produced a steep relief.

The only outlet for Gulf waters occurs in the southeast, where the island of Cuba partially plugs the gap between the Yucatán and Florida peninsulas. Most outflow from the Gulf is into the Caribbean through the Yucatan Channel; a minor drainage into the Atlantic is through the shallow Straits of Florida.

The arc of the northwest Florida coastline from Cedar Key in Levy County to Alligator Point in Franklin County faces a shallow sea that extends for a considerable distance offshore with little or no appreciable increase in depth. During periods of extremely low tides, the bay floor may be exposed for a half mile (0.8 km) or more, except along channel courses. These mud flats, as they are called, are partially attributable to an underlying bedrock of resistant limestone that continues landward, outcropping occasionally along the shore. Little sand is brought down by mainland rivers that pass through the limestone deposits, and the small amount available encounters restrained wave action due to the long fetch across the shallow bottom. These features, together with weak wind action and the absence of prevailing longshore currents, result in a coastline with few beaches. There are extensive salt and brackish marshes receiving nourishment from normal tidal flow through inlets, from freshwater streams, and from seasonal storm tides that breach the sand ridges leaving water trapped behind. Low flatwoods of slash pine and palmetto originate where the marshes feather out.

The sandy flange of this section of the Gulf coast is moving landward, as indicated by the occurrence of marsh peat beneath barrier beaches and offshore bars, but the rate of encroachment is exceedingly slow, apparently due to the low level of energy. St. Mark's Lighthouse, situated now at the water's edge, was placed on its present foundation in 1842.

Fishing is the principal attraction of this section of the coast, and the names of many bayside communities are synonymous with this sports activity. Many of the wetlands, shallow bays, and sluggish channels ringing Apalachee Bay are Florida Wildlife Management Areas or National Wildlife Refuges, affording protection for migratory waterbirds as well as other fauna and the flora.

At Panacea the Ochlockonee and Sopchoppy Rivers converge to form Ochlockonee Bay. Alligator Point, across the bay from Panacea, is a long spit that narrows to a mere dune at places, expanding into pine-oak woods at others. Alligator Harbor is located where the spit doubles on itself, forming a hook. At the wide base of the spit there are broad, flat, old dunes of fine, white sand, completely bare in spots, with lichens forming a sparse groundcover elsewhere. Rosemary (*Ceratiola ericoides*) and gopher-apple (*Licania michauxii*) are present in profusion. Blueberry (*Vaccinium myrsinites*) and dune greenbrier (*Smilax auriculata*) are here, too, together with bull-nettle (*Cnidoscolus stimulosus*), and a variety of seasonal wildflowers. Unfortunately, this striking landscape may soon be obliterated if present residential development expands to include it.

Alligator Point to the Mississippi Delta

Streams entering the Gulf between Alligator Point and the Mississippi Delta pass through sandy clay formations rather than limestone, bringing sand and silt that, through the ages, have formed magnificent beaches and dunes, both on the mainland and the islands. The coastline is marked, except for the Apalachicola Delta, by a succession of embayments that are the drowned valleys of current or former rivers. Mobile Bay is the largest of these and represents the major drainage system for Alabama. Pensacola, Choctawhatchee, and St. Andrew Bays are major embayments in northwest Florida, each with elongate spits and islands narrowing their entrances. The principal drainage systems in Mississippi, discharging into Mississippi Sound, are Pascagoula, Pearl, Biloxi, and Bay St. Louis. Of these four, Pearl and Pascagoula are presently active rivers.

The coasts of northwest Florida and eastern Alabama are flanked by narrow islands, spits, and bars lying fairly close to shore. Dauphin Island, Alabama, and a chain of islands in Mississippi stand at a distance of some 6 to 12 miles (10–19 km) offshore, separating Mississippi Sound from the open Gulf. These islands are generally alike in that they are long, low, and sandy; they have beaches, dunes, marshes, and sometimes forests. Unlike barrier islands of the Atlantic, the landward side is not normally bordered by lagoons and continuous marshes. This is an active hurricane zone and storm overwash is frequent.

Northwest Florida's easily accessible, wide, white sandy beaches have made it a popular resort area for many years, but current expansion of facilities appears to be of gigantic proportions. It is possibly too late for concern over dune destruction, eradication of vegetation, and general pollution, although Gulf Islands National Seashore on Santa Rosa Island and a few small state parks may preserve a semblance of the area as it once looked. Dauphin Island, Alabama, is owned in parts by local, state, and federal governments and private interests. It is heavily developed and the bridge linking it to the mainland, destroyed in a recent hurricane, has been replaced. In Mississippi, Petit Bois, Horn, and Ship Islands are part of the Gulf Islands National Seashore; Cat Island is privately owned. In Louisiana, the Chandeleur Islands are a National Wildlife Refuge and large sections of the Mississippi Delta are wildlife refuges or management areas.

In the state of Mississippi, the past few decades have witnessed extensive destruction for residential construction, commercial sites, shipping channels, garbage disposal, and similar activities. Eleuterius (1972) wrote that projected plans for the state to be executed by 1990 will bring to 52% the total of Mississippi marsh destroyed by development.

There are along the coast, however, a few vestigial spots that remain in a reasonably unspoiled state where one may go for reminders of the original beauty once found throughout the area. Santa Rosa Island is one, a narrow spit formation lying between the Gulf and Santa Rosa Sound near the mouth of Pensacola Bay. It has a fine beach of white sand and a large, active primary dune system with a fair covering of sea-oats. A short distance behind the foredunes and surprisingly close to open Gulf waters, isolated stands of magnolia, atypically low in height but spreading broadly across the dunes, bloom handsomely in early April. Woody vegetation of the secondary dune field is generally sparse, but there is still the profile, once typical of the coast as a whole, of wind-sheared red-bay, scrub oak, and sand live oak (*Quercus geminata*). *Asclepias humistrata*, the colorful sandhill milkweed, and bull-nettle with its showy milk-white sepals are part of the diverse flora that occurs here.

Mississippi Delta to the Sabine River

The delta of the Mississippi River dominates the northern coast of the Gulf of Mexico. Bolander et al. (1970), in a study of the lower Mississippi Delta, reported that this dynamic stream drains roughly 60% of the conterminous United States, transporting an estimated sediment load of two million tons daily, borne by an incredible volume of water. The delta that receives and is partially formed by this load is an extensive body with a depth up to 13,000 feet (4,000 m). It is produced by fluvial and marine deposits and is dissected by numerous distributaries. The present delta is the latest of a series of deltas, each formed when the river coursed a different pattern, the whole or total accounting for the massive protuberance extending into the Gulf. The inactive delta systems have created a lowland honeycombed with lakes, bayous, and marshes, some brackish and some freshwater, extending as far west as the Sabine River.

Mississippi waters entering the Gulf flow westward as a vigorous longshore current and account for the formation of numerous barrier islands. These islands are constantly shifting, but characteristically have sandy beaches and low dunefields on the seaward side and wetlands on the inner side, either black mangrove swamps or marshes composed principally of smooth cordgrass. Marsh islands produced by old waterways within the delta itself are frequently freshwater or only slightly brackish and are vegetated by grasses, forbs, shrubs, and trees. Artificial diking has been employed to assure fresh water in some areas for rice cultivation, and southwest Louisiana is one of the largest rice-producing areas in the United States. Sugarcane is another commercial crop grown successfully in the delta.

Transportation along most of the Louisiana coast is best effected by boat, as the terrain is not conducive to road construction. Several beaches in Cameron Parish, however, have small communities of beach houses that can be reached by car and Grand Isle in Jefferson Parish, a state park, is also accessible by car. Vast sections of the delta have been designated wildlife refuges, both state and national, and offer a rich resource for studying native flora and fauna, especially waterfowl.

Literature Cited

Bolander, R. J., T. M. Gard, A. G. Richtmyer, and C. C. Almy. 1970. *A study of the lower Mississippi River delta, its processes, sediments, and structures.* Geological Society of New Orleans field trip guide book. 48 p.

Correll, D. S., and M. C. Johnston. 1970. *Manual of the vascular plants of Texas.* Texas Research Foundation, Renner. 1881 p.

Eleuterius, L. N. 1972. The marshes of Mississippi. *Castanea* 37: 153–168.

Fernald, M. L. 1950. *Gray's manual of botany,* 8th ed. American Book Co., New York. 1632 p.

Gleason, H. A., and A. Cronquist. 1963. *Manual of vascular plants of northeastern United States and adjacent Canada.* D. Van Nostrand Co., Inc., Princeton, NJ. 810 p.

Godfrey, R. K., and J. W. Wooten. 1981. *Aquatic and wetland plants of southeastern United States.* 2 vols. University of Georgia Press, Athens. Vol. Dicotyledons, 933 p.; vol. Monocotyledons, 712 p.

Gosselink, J. G., E. P. Odum, and R. M. Pope. 1973. *The value of the tidal marsh.* Center for Wetland Resources, Louisiana State University, Baton Rouge. 25 p.

Hoyt, J. H., and V. J. Henry, Jr. 1971. Origin of capes and shoals along the southeastern coast of the United States. *Geol. Soc. Am. Bull.* 82: 59–66.

Radford, A. E., H. E. Ahles, and C. R. Bell. 1964. *Manual of the vascular flora of the Carolinas.* University of North Carolina Press, Chapel Hill. 1183 p.

Reader's Guide

We hope that this book will appeal to both amateurs and scientists. Because much of the information is, by necessity, technical, we have included a number of aids that will enable easier usage by the amateur botanist. These include an extensive glossary; six plates showing vegetative, floral, and fruit characteristics mentioned in the descriptions and keys; 186 line drawings of distinctive characters accompanying the descriptions; and almost 600 color photographs. Species descriptions and corresponding photographs have been numbered to allow cross-reference from text to picture and from picture to text.

Amateurs may simply ignore technical aspects of this book and, because of the easily understood material and the photographs and drawings, will be as well off as they would be with a book lacking the technical parts. This book has the advantage of allowing the amateur to become an expert.

Method of Presentation

Standard floras include all vascular plants for a stated area and present keys and descriptions for their identification. This book differs in that it covers only seed-bearing plants of the coastal zone, and some of the less common species are not included. Color photographs of 588 coastal species are presented, accompanied by descriptions of diagnostic features not present or readily apparent in the photographs or that need emphasis, both ecological and geographical distributions along the coast, and flowering dates. Where applicable, additional species similar to the ones illustrated are included in the descriptions, with distinguishing characteristics for their recognition. A glossary is provided on p. 47.

The partially artificial system of species arrangement employed is designed to facilitate rapid use of the book. Major separations are made between woody plants; grasslike plants such as grasses, sedges, and rushes; and the remaining herbaceous, nongrasslike plants known as forbs. Within each major unit, family arrangement is followed, as characteristics of many families are easy to learn, facilitating identifications.

Woody plants are categorized here as those in which some or all above-ground stems live from one year to the next. Differentiation between tree and shrub is arbitrary and either form may occur within a given species. Foresters and horticulturists base the distinction on height of the plant and diameter of the trunk at a given height, or whether there is, in fact, a main trunk. A tree, by our definition, has a perennial stem with a height of at least 13 ft (4 m) and a minimum diameter of three inches (7.5 cm) at breast height. Any other woody plant is designated a shrub, vine, or chamaephyte.

Deviations obviously occur and all the more frequently in circumstances of environmental stress as found along seacoasts. Some trees and shrubs that attain considerable height inland may hover near ground level where subjected to gale winds. Conversely, species normally thought of as shrubs may reach tree dimensions under unusually favorable circumstances. Beach plum is normally a shrub of three to six ft (one or two m) height, but can mature and bear fruit at a maximum height of six inches (15 cm), or it may attain tree size. The amateur is easily led into identifying as herbaceous several of the chamaephytes because they are small and the woody stems not always noticed.

Grasses and grasslike plants such as rushes and sedges exhibit graceful forms that enhance the landscape and merit attention. Many are included here not only for their prominent role in the coastal community, but because they are often overlooked and much beauty goes unobserved. They are grouped together because of their broad vegetative similarities such as narrow, linear leaf blades.

Recommendations for Identification

To use this book effectively for the identification of a given plant, an orderly procedure of elimination is recommended.

(1) First, determine if the plant is woody; that is, does a conspicuous portion of the stem survive and persist over the winter? If so, turn immediately to Key to Groups of Woody Plants . . . . p. 64

(2) If the plant is not woody, closely examine the plant structure and, on the basis of your observations, find its appropriate place in the Key to Species Groupings . . . . p. 35

(3) Use the appropriate Key, as directed in the Key to Species Groupings in (2) above. This could lead directly to a species or genus, but more likely to a family.

(4) Finally, check the plant with the photographs and any line drawings for similarities. If these match sufficiently, compare descriptions and distributions.

(5) As a short cut, reference to Plants with Distinctive Characteristics (p. 35) might result in a quick identification of some species.

The most valuable skill in plant identification is the ability to detect differences between species. Particular attention must be paid to numbers, shapes, and arrangements of plant parts, especially of flowers and fruits. One must know what constitutes a single flower, for example, and check for heads of flowers as in the Asteraceae and some Fabaceae. Use of the plates beginning on page 41 will be helpful at this juncture. These illustrations are grouped for convenience under three major headings: vegetative, floral, and fruit characteristics.

If the above process should prove unsuccessful, it is hoped the user will try again, following the same procedures, checking for errors in observation. This should eventually lead to success, as essentially all species likely to be encountered in seaside habitats are included in this book.

Key to Species Groupings

Species included in this book are divided into four major groupings, three herbaceous and one woody. Keys to groups of woody plants are presented last (on pp. 64–71); the three herbaceous groupings are in the following order with the first subdivided into two parts:

Leaves net-veined	Leaves parallel-veined	Leaves parallel-veined
Sepals and Petals 4–5 each, sometimes many, usually evident	Sepals and Petals 3 each, usually evident	Sepals and Petals absent or not evident due to a form different from that ordinarily recognized
All are dicots	All are monocots	All are monocots
Petals separate / Petals united	See Key III (p. 59)	See Key IV (p. 61)
See Key I (p. 56) / See Key II (p. 58)		

Plants with Distinctive Characteristics

Some groups of plants are set apart from others by distinctive characteristics, providing an advantage for purposes of identification. Some of these characteristics and the families or genera exhibiting them are given below. These features are not necessarily confined to the plant groups mentioned, nor do all species in each group necessarily conform to all the characteristics listed, but in general the list assists the user in identifying plants rapidly.

1. Flowers usually small and packed tightly in heads: Fabaceae (pp. 249, 369), Asteraceae (pp. 294, 389), *Eriocaulon* (p. 313), *Phyla* (p. 283), *Xyris* (p. 313).
2. Flowers in simple umbels: Liliaceae (p. 317), Asclepiadaceae (p. 278), *Centella* (p. 272), *Hydrocotyle* (p. 272).
3. Flowers in compound umbels: Apiaceae (p. 272).
4. Sepals and petals 3 each and all similar: Liliaceae (p. 317), Amaryllidaceae (p. 318), Iridaceae (p. 318), *Juncus* (pp. 315–17).
5. Stamens numerous: Ranunculaceae (p. 244), Rosaceae (pp. 248, 364), Hypericaceae (pp. 262, 376), Malvaceae (p. 261), Alismataceae (p. 312), Nymphaeaceae (p. 243).
6. Plants lacking green color: *Cuscuta* (p. 280), *Corallorhiza* (p. 321), *Monotropa* (p. 275).
7. Plants with milky juice: Asclepiadaceae (p. 278), Apocynaceae (p. 278),

Euphorbiaceae (p. 259), Asteraceae (p. 294).

8. Petals united into a tube and with a distinct upper and lower lip: Lamiaceae (pp. 283, 387), Scrophulariaceae (p. 287), *Lobelia* (p. 294).

9. Flowers with 4 separate sepals, 4 separate petals, and 6 stamens (2 short, 4 longer): Brassicaceae (p. 245).

10. Leaves thick, succulent, smooth: *Batis* (p. 362), *Honkenya* (p. 242), *Portulaca* (p. 240), *Sesuvium* (p. 240).

11. Inflorescences narrow and coiled backward at tips: *Heliotropium* (p. 281).

12. Inflorescences apparently attached between nodes: Solanaceae (p. 286), *Phytolacca* (p. 239).

13. Leaves floating on water: Nymphaeaceae (p. 243), *Brasenia* (p. 243), *Nymphoides* (p. 277).

14. Leaves peltate: *Brasenia* (p. 243), *Hydrocotyle* (p. 272).

Comments on Photographs and Drawings

Representative plants or parts of them were photographed for each species illustrated in this book. Sizes of the images were recorded from the camera indicator, or computed from measurements of live plant and photographic images or from comparable measurements of images and pressed dried specimens. The relative size of the pictures in the book is indicated by a number placed after the scientific name in the captions of the photographs. For example, × 2 means the picture is twice the actual size, × 1 means the sizes are the same, and × ¼ means the picture is one-fourth the size of the actual subject.

Positive identity was made possible by collecting and preserving samples of the plants photographed and checking them against herbarium specimens and in manuals and other literature. Plants with which we were quite familiar were sometimes not collected.

Drawings are included for many species to illustrate important characteristics not shown in the photographs. For example, closeup photographs including flowers usually do not show midstem leaves and often do not include fruits, which appear later. Some drawings illustrate variations that occur in shape of leaves or other plant parts too diverse for a single photograph.

Comments on Descriptions

The description gives for each species its common and scientific names, recognition characteristics, abundance, ecological and geographical distribution, flowering period, and often other scientific names previously used (synonyms).

Plant Names. Scientific names we have used are, for the most part, those employed by technical manuals. Some names differ from those in manuals because of changes indicated by current literature. In the event of a name change, the scientific name used in the manual is listed as a synonym.

The common names chosen are the ones most often applied to the species. However, common names are frequently unreliable, as they can vary between localities or even within a single community, leading to confusion and misunderstanding. Common names do have interest and value and are frequently the only means of communication for those people unfamiliar with scientific names.

Recognition Characteristics. The description for each illustrated species provides additional characteristics usually not present or readily apparent in the photograph, but necessary for distinguishing the species from similar ones occurring in seaside habitats. In using the descriptions, reference to the illustrations in the plates on pp. 41–46, and to the Glossary (p. 47), where terms are described as simply as possible, will be helpful, especially to amateurs.

If more than one species of a genus is illustrated, some of the characteristics common to all may be given with the first species presented. It should become a matter of routine to check the first species description for these general characteristics. Similarly, descriptions for the first species included for a family should be checked for characteristics common to all seaside species of that family.

Measurements. The metric system is used in this book. By no other practical method can sizes of the small parts of flowers be compared. For your convenience a metric scale 20 centimeters (cm) long has been placed inside each cover. Twenty cm equals 200 millimeters (mm). A meter is 5 times that long (i.e., 100 cm, 1000 mm, or 39.4 inches). Ten mm equals 1 cm.

Abundance. A subjective evaluation of abundance for each species has been made on the basis of our field experience, comments of other specialists, and analysis of scientific studies and specimens preserved in herbaria.

Abundance of a species is given as common, occasional, or rare. Because these are subjective designations, only common and rare should be considered as significantly different ratings. Any species listed as occasional, however, is neither quite rare nor common. Species that have limited distribution, although they may be abundant where found, are listed as rare, as are those that have wide distribution but are represented by few plants numerically. Species listed as common may be occasional, rare, or even absent in a given locale.

Distribution. The natural occurrence of a species may be considered in two categories: (1) the kind of habitat(s) in which it grows; and (2) its geographical distribution, also called range. Different habitats typically result from variations in light, moisture, soils, and associated species. Knowing that a species usually occurs in a certain habitat can be helpful in identifying a specimen or in finding plants of a given species. Geographical distributions are similarly useful, and for these reasons, as well as for general interest, the kinds of habitats and ranges are given for each species in this book.

Ranges are denoted for seaside habitats only and indicated by states from Texas to Maine and by provinces in Canada. Occurrences in Mexico and all inland distributions are omitted and, as pointed out in the introduction, most of peninsular Florida is excluded. Standard United States Postal Service abbreviations for states are used. Those used for Canadian provinces are as follows: New Brunswick, NBr; Nova Scotia, NS; and Newfoundland, Nfld. The designation for Greenland is Greenl (see p. 39).

Distributions were determined from our field observations; the many published studies involving seaside habitats; studies by others of families, genera, and groups of species; examination of specimens in herbaria; and, with certain reservations, from manuals. Distribution for a portion of the seaside range for a state or province is indicated by "w" or "e" for states from Texas to Florida, and by "s" and "n" along the Atlantic coast from Georgia to Newfoundland. For example, a range given as eLA to sVA means that the species is known to occur in seaside habitats in all or part of the eastern half of Louisiana, in all or part of the southern half of Virginia, and in the states between them. If the species is known to occur somewhere in the northern half of Virginia, then the distribution is indicated as eLA to VA. Occasionally, and especially in the case of rare species, a species may be absent from one or more of the states included in the range. This may be indicated by a break, as for *Drosera capillaris:* TX–FL; NC–Nfld.

Flowering Period. The period of flowering is given by month(s) for each species for its entire seaside distribution from Louisiana to Massachusetts. In a flowering period indicated as Mar.–June for a species present throughout this entire range, however, the species is unlikely to be found in flower in March in the northern states, namely NJ, NY, CT, RI, and MA, and may not be found in flower in June in the Gulf states.

Other Scientific Names. Botanists do not always agree on which scientific names to use. Some of the scientific names found in other books may not be used here. Such differences in nomenclature could be due to any of a number of technical reasons: an earlier name being found, the species being better placed in another genus, etc. Names used by others but not by us, for whatever reasons, are occasionally listed after the flowering period, as synonyms. This is done especially when the names will help the reader find the species in other books.

Other Similar Species. Many plants cannot be reliably identified to species from photographs in wildflower books, and this book is no exception. A special effort has been made, however, to include in the descriptions characteristics necessary for identification. Providing these characteristics also makes it reasonably easy to identify other similar species with one or more contrasting characteristics. For many species this extra information has been provided in the final paragraph(s) of the descriptions. Common names are generally omitted for these extra species because they are usually the same as for the ones illustrated. Abundance, distribution, and flowering period ordinarily are given. If omitted, they are essentially the same as that in the previous paragraph.

Abbreviations

Geographic Directions
e - east
w - west
n - north
s - south

States
TX - Texas
LA - Louisiana
MS - Mississippi
AL - Alabama
GA - Georgia
FL - Florida
SC - South Carolina
NC - North Carolina
VA - Virginia
MD - Maryland
DE - Delaware
NJ - New Jersey
NY - New York
CT - Connecticut
RI - Rhode Island
MA - Massachusetts
NH - New Hampshire
ME - Maine

Metric Scale
mm - millimeter
cm - centimeter
m - meter

Provinces
NBr - New Brunswick
NS - Nova Scotia
Nfld - Newfoundland
Greenl - Greenland

Plates
Vegetative, Floral, and Fruit Characteristics

VEGETATIVE STRUCTURES

SHAPE OF BLADE (LEAF, PETAL, SEPAL, BRACT)

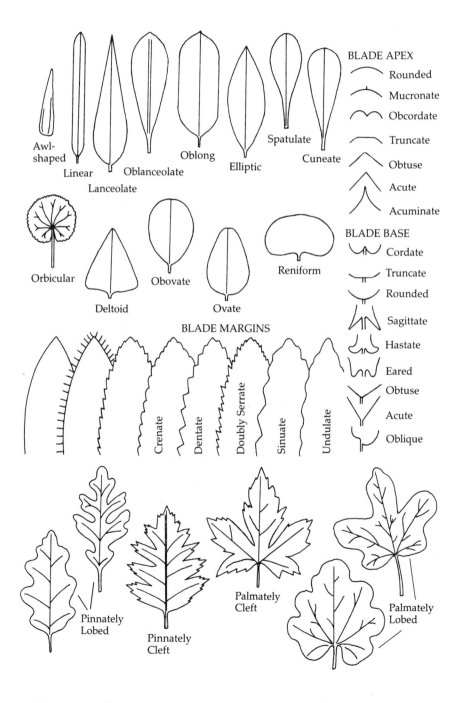

VEGETATIVE STRUCTURES

SIMPLE LEAVES

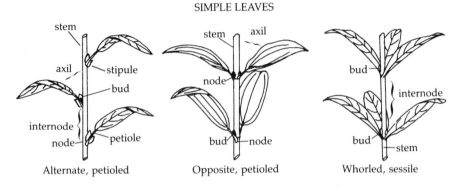

Alternate, petioled

Opposite, petioled

Whorled, sessile

ONCE COMPOUND LEAVES

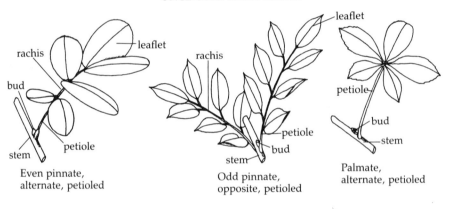

Even pinnate,
alternate, petioled

Odd pinnate,
opposite, petioled

Palmate,
alternate, petioled

TWICE COMPOUND LEAF

OTHER LEAF TYPES

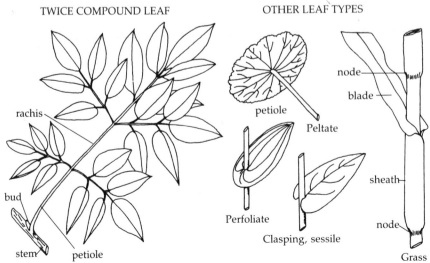

Peltate

Perfoliate

Clasping, sessile

Grass

FLORAL STRUCTURES

COMPLETE FLOWERS WITH SUPERIOR OVARY AND NO HYPANTHIUM

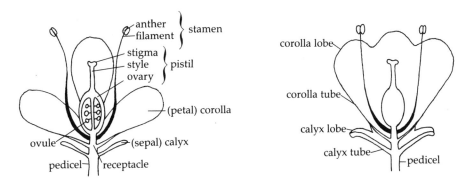

COMPLETE FLOWERS WITH A HYPANTHIUM (H)

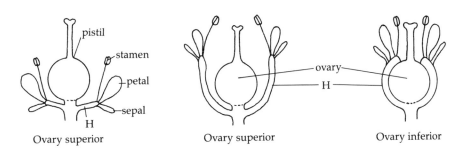

Ovary superior   Ovary superior   Ovary inferior

FLOWER FORM

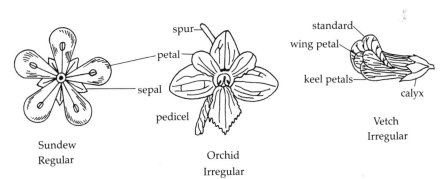

Sundew
Regular

Orchid
Irregular

Vetch
Irregular

FLOWER OR FRUIT CLUSTERS

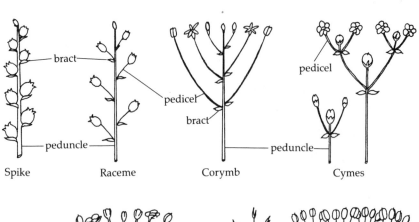

Spike   Raceme   Corymb   Cymes

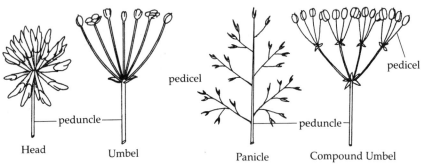

Head   Umbel   Panicle   Compound Umbel

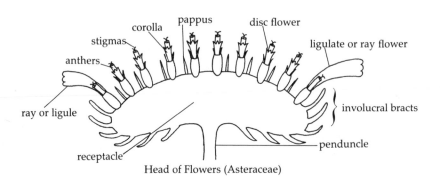

Head of Flowers (Asteraceae)

PLACENTATION TYPES

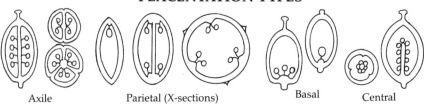

Axile   Parietal (X-sections)   Basal   Central

DRY FRUIT TYPES

(Interior or Surface Views)

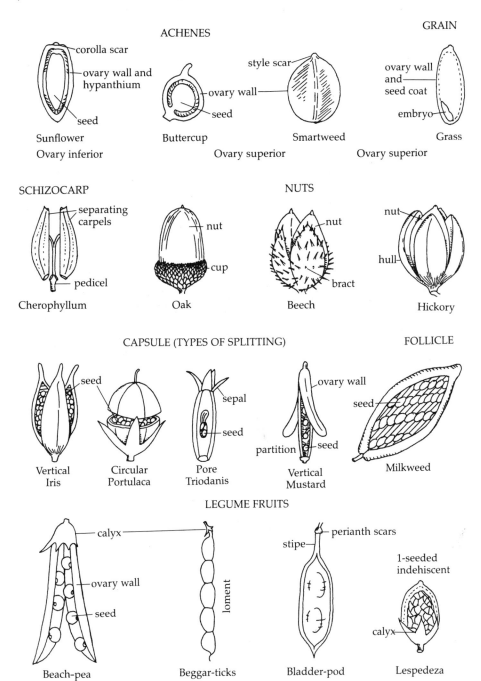

ACHENES

corolla scar
ovary wall and hypanthium
seed

Sunflower
Ovary inferior

ovary wall
seed

Buttercup

style scar
ovary wall

Smartweed
Ovary superior

GRAIN

ovary wall and seed coat
embryo

Grass
Ovary superior

SCHIZOCARP

separating carpels
pedicel

Cherophyllum

nut
cup

Oak

NUTS

nut
bract

Beech

nut
hull

Hickory

CAPSULE (TYPES OF SPLITTING)

seed

Vertical
Iris

Circular
Portulaca

sepal
seed

Pore
Triodanis

ovary wall
partition
seed

Vertical
Mustard

FOLLICLE

seed

Milkweed

LEGUME FRUITS

calyx
ovary wall
seed

Beach-pea

loment

Beggar-ticks

stipe
perianth scars

Bladder-pod

1-seeded indehiscent

calyx

Lespedeza

FLESHY FRUIT TYPES

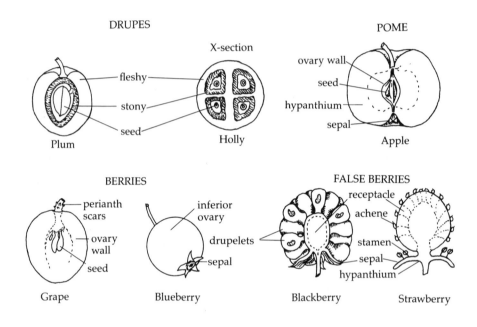

DRUPES

X-section

POME

fleshy

stony

seed

Plum

Holly

ovary wall

seed

hypanthium

sepal

Apple

BERRIES

perianth scars

ovary wall

seed

Grape

inferior ovary

drupelets

sepal

Blueberry

FALSE BERRIES

receptacle

achene

stamen

sepal

hypanthium

Blackberry

Strawberry

Glossary

ACCRETION   Increase in size by gradual addition on the outside.

ACHENE   A small, dry, one-seeded, indehiscent fruit (as in Asteraceae, *Ranunculus*, and *Polygonum*); distinguished from a nutlet by its softer outer wall. These types intergrade.

ACUMINATE   Tapering gradually to a long, thin point. Compare with acute.

ACUTE   Applied to tips and bases of structures ending in a sharp point less than a right angle. Compare with acuminate and obtuse.

ALTERNATE   One leaf, bud, or branch per node.

ANNUAL   Plant growing from seed to fruit in one growing season, then dying naturally.

ANTHER   The pollen-bearing part of a stamen.

ANTRORSE   In a forward and upward direction. The opposite of retrorse.

APPRESSED   Lying flat against.

ASCENDING   Rising obliquely or curved upward.

AURICLE   An ear-shaped appendage or projection. In pairs at base of some leaf blades and petals.

AWL-SHAPED   Having a linear shape and tapering to a fine point; narrowly triangular.

AWN   A bristle-shaped structure.

AXIL   The space between any 2 adjoining organs, such as stem and leaf.

AXILE PLACENTA   See PLACENTA.

AXILLARY   In an axil.

BAR   Offshore ridge of sand or sediment submerged by high tide and exposed at low tide.

BARB   A sharp projection extending backward, as in a fishhook.

BARRIER ISLAND   Offshore body of land that rises permanently above normal sea level.

BARRIER SPIT   Same as barrier island except still attached to parent body of land at one end.

BASAL PLACENTA   See PLACENTA.

BEACH   That portion of the shore lying between lowest low water and highest high water and composed of loose sediments. These sediments consist of sand and sometimes shell hash along southern shores, sand or shingle (gravel, pebble, or cobble) along northern shores.

BEACH RIDGE   Elongate mound of coarse sand or gravel formed by wave action along upper margin of beach.

BEAKED   Ending in a firm, prolonged, slender tip.

BERM   Backshore terrace lying between beach and dunes.

BERRY   Any fruit with fleshy walls and with a few to many seeds encased in soft tissue, such as a grape, tomato, or pepper. See DRUPE.

BIENNIAL   Plant growing from seed to fruit in 2 growing seasons, then dying naturally.

BIGHT   Bend in a coast forming an open bay; also the bay so formed.

BISEXUAL   Having stamens and pistils present and functioning in the same flower.

BLADE   The flattened and expanded part of a leaf, or parts of a compound leaf.

BLOWOUT   Break in dune formation due to wind or wave action, usually at a weakened spot.

BRACT   A reduced leaf, particularly at bases of flower stalks and on the outer part of the heads of flowers in the Asteraceae.

BULB   A swollen structure composed of concentric layers of modified leaves, as in an onion; usually subterranean.

BUNDLE SCAR   Scar within a leaf scar where vascular bundle broke off when leaf detached.

BUR   A rough, prickly, or spiny flower or fruit (or cluster of these) and any associated parts.

CALYX   Collective term for all the sepals of a flower, whether separate or united (i.e., the outer series of flower parts).

CAPE   Piece of land or promontory, relatively extensive in size, jutting into the sea from a major land mass and prominently altering the coastal profile.

CAPILLARY   Hairlike.

CAPSULE   A dry fruit with 2 or more rows of seeds and splitting open (dehiscing) at maturity.

CARPEL   A simple pistil, or unit of a compound pistil.

CHAMAEPHYTE   A low shrub (subshrub); e.g., *Hudsonia*.

CILIATE   Fringed with hairs. Can be said of any hairy margin, whether entire, toothed, lobed, or otherwise.

CIRCUMBOREAL   Around the world in northern regions.

CLAW   The narrowed parallel-sided base of the petals in some kinds of flowers; e.g., *Silene*.

CLEFT   Having deeply cut lobes or parts; usually divided half way to the midvein or more.

CLIMAX   Relatively stable vegetation in equilibrium with its environment.

COAST   Term for general area falling under influence of the sea.

CONIC   Cone-shaped.

CORIACEOUS   Stiff, leathery texture.

COROLLA   Collective term for all the petals of a flower, whether separate or united (i.e., the inner series of the perianth).

CORYMB   An approximately flat-topped inflorescence in which the flower stalks arise from various points of the main stem and in which the opening of flowers begins at the outer edge and progresses inward.

CRENATE   Having margin scalloped with shallow, rounded teeth or lobes.

CUNEATE   Triangular or wedge-shaped with narrow end at point of attachment to another plant part.

CUTICLE   Waxy, noncellular layer on outer surface of epidermal wall of a plant.

CUTIN   Waxy, noncellular material coating the outer surface of the epidermal wall of a plant; forms cuticle.

CYATHIUM(A)   The basic inflorescence type of *Euphorbia*. Consists of cuplike involucre bearing flowers from its inner base.

CYME   A type of inflorescence that is approximately flat-topped and in which the main axis is always terminated by a single flower that opens before those lateral to it.

DECLINED   Bent downward or drooping.

DECUMBENT   Lying on the ground, the far end erect or ascending.

DECURRENT   Extending down and attached to the stem, forming a ridge or wing.

DEFLEXED   Bent or turned abruptly downward.

DEHISCENT   Opening by natural splitting, as an anther discharging pollen or a fruit its seeds.

DELTOID   Triangular in shape.

DENTATE   Toothed; having sharp, coarse, conical projections perpendicular to the margin. Compare with serrate.

DETRITUS   Accumulated mass of disintegrated material deposited primarily by water or wind, or both.

DIAPHRAGM   A cross-partition; a membrane that separates.

DICHOTOMOUS   Branching by successive forking, always into 2 approximOUely equal divisions.

DICOT   A major group of flowering plants generally characterized by net-veined leaves, flower parts 4–5 or multiples thereof, and vascular bundles of stem in a circle.

DISC (DISK) FLOWER   Flower with tubular corolla and lacking ligule (ray) in heads of some Asteraceae.

DIVERGENT   Inclined away from each other.

DRIFT   Fragments of detritus or flotsam deposited along high water line.

DRUPE   Fleshy indehiscent fruit, such as cherry or plum, having a single seed encased in a hard stony covering, or sometimes with more than one such seed, as in holly.

DUNE   Pile or ridge of loose sand deposited by wind action.

DUNE, PRIMARY   Continuous dune ridge lying nearest the ocean and running generally parallel to the shoreline.

DUNE, SECONDARY OR STABLE   Dune system lying landward of and older than the primary dune, stabilized by vegetation.

DUNELET   A small dune.

ELLIPSOID   Said of a 3-dimensional body whose plane sections are all either ellipses or circles.

ELLIPTIC   Oblong with the ends equally rounded or nearly so. Not as wide as oval.

ENTIRE   Smooth, without teeth or indentations; applied to margins, edges.

EPIPHYTE   A plant with no connection to the soil, growing upon another plant, but deriving no food or water from it; not parasitic.

ERICAD OR ERICACEOUS   Belonging to the heath family of plants (Ericaceae), a large group that includes blueberries, cranberries, and rhododendrons.

ESTUARY   Mouth of a river where it joins the sea and where fresh water and salt water mix.

EVERGREEN   Plants with live leaves persisting through one or more dormant winter seasons.

FALCATE   Crescent-shaped.

FASCICLED   Borne in a bundle, clustered.

FEMALE FLOWER   With pistils but no fertile stamens.

FILAMENT   The part of a stamen to which the anther is attached; the stalk of a stamen; usually slender, thread-like.

FILIFORM   Theadlike; long and very slender.

FLAT, SALT   Level area in a high salt marsh that fails to drain fully at low tide; evaporation results in an accumulation of salts in the soil. Also called pan or panne.

FLAT, SAND   Interior behind or between dunes that is comparatively level, open, and with low vegetation.

FLATWOODS   Poorly drained, low-lying, generally level timberland.

FLORET   A small flower that is an individual component of a dense inflorescence, as in grasses and composites.

FOLLICLE   A dry one-celled fruit with a marginal placenta and splitting along one edge, such as the pod of milkweed.

FORB   Any herbaceous plant other than a grass, sedge, or rush.

FRUIT   A ripened ovary, together with the parts regularly attached to it (i.e., the seed-bearing structure of the plant and any attached parts).

GENICULATE   Bent abruptly at an angle, like a bent knee.

GLABROUS   Without hairs, bristles, or stalked glands.

GLAND   A secreting surface, appendage, or protuberance or any structure resembling such an organ.

GLANDULAR   Bearing glands; these may be sessile, on short stalks, on tips of hairs, or sunken in leaves or other structures.

GLAUCOUS   Whitened, as on a surface with a fine white substance (bloom) that will rub off, as on many grapes and blueberries.

GLOBOSE   Globe-shaped; spherical.

GLOMERULE   Compact cluster of 2 or more flowers.

GLUME   Each of the 2 empty chaff-like bracts at the base of grass spikelets. Occasionally only one.

GRAIN   The small, hard, indehiscent, seedlike fruit of the grass family; the seed is fused to the ovary wall.

GRANULE   Minute grain.

GROIN   Wall or structure built out into the surf, away from inlets, for beach protection against erosion.

GROUND WATER   All water below the surface of the water table.

HALOPHYTE   Plant tolerant of various salts, notably sodium chloride (NaCl).

HAMMOCK   A raised fertile area in the midst of a wetland, characterized by hardwood vegetation and deep humus soil.

HASH   Shell fragments produced by water action and weathering.

HASTATE   Said of arrow-shaped leaves with narrow basal lobes that spread outward.

HEATH   An open, rather level area with poor drainage, poor soil, and a surface rich in peat or peaty humus.

HERBACEOUS   Not woody; dying down each year.

HERBACEOUS PLANT   A plant with no persistent woody stem above ground (i.e., dying back to the ground at end of growing season).

HIP   The fleshy to leathery hollow fruit of roses, developed from the hypanthium, and with achenes attached to the inside.

HOLOCENE   More recent of 2 Epochs of the Quaternary Period of geologic time that began 1.5 million years ago. See PLEISTOCENE.

HOOK   Spit of land that curves inward toward land mass to which it is attached.

HURRICANE   Tropical cyclone or intense low-pressure system that generates winds of over 75 miles (120 km) per hour.

HYALINE   Transparent or translucent.

HYPANTHIUM   A saucer-shaped, cup-shaped, or tubular structure below, around, or adhering to the sides of the ovary. Sepals, petals, and stamens are attached at or near the outer or upper margin of the hypanthium.

IMBRICATE   Overlapping, as with shingles.

INDEHISCENT   Not opening naturally at maturity.

INFERIOR   Descriptive of an ovary fused to the hypanthium, the ovary thus appearing to be located below the sepals and petals.

INFLORESCENCE   Any complete flower cluster on a plant including branches and bracts. Clusters separated by leaves are separate inflorescences.

INTERNODE   That portion of the stem between 2 adjacent nodes.

INVOLUCRE   A set of bracts below or around a single flower or head of flowers, as in the Asteraceae.

IRREGULAR   Said of flowers with sepals or petals of unequal size or shape or both.

JETTY   Wall or similar structure built out into the surf from the shore to prevent transport of sediments into harbors and inlets.

KEEL   A central ridge on back of plant parts such as sepals, petals, bud scales, glumes. The 2 lower adhering petals of some legume flowers.

LAGOON   Body of water between a barrier island, spit, or bar and the mainland. Also may be called sound, bay, river, creek, lake.

LEAF SCAR   Mark left on a twig where a leaf has broken off.

LEAFLET   A single segment (blade) of a compound leaf.

LEGUME   A dry one-celled fruit with a marginal placenta and usually splitting along the 2 edges; specifically the fruit of the Fabaceae.

LEMMA   The outer, lower bract subtending the floret in a grass spikelet.

LENTICEL   Small corky spot or line on the young bark of stem or trunk of a woody plant.

LIGULE (RAY)   Strap-shaped part of the corolla in many of the Asteraceae. Most often on the marginal flowers of the head as in the sunflower; but all flowers may have ligules, as in dandelion.

LINEAR   Narrow and elongated with sides parallel or nearly so.

LIP   The upper or lower part of an unequally divided corolla or calyx.

LITTORAL   Pertaining to a shore, especially of the sea; zone along coast between tides affected by waves and coastal currents.

LOMENT   A legume composed of one-seeded sections that separate transversely into joints; e.g., the fruit of *Desmodium.*

MALE FLOWER   With stamens but no functional pistil.

MARSH   Land area constantly saturated or flooded with water, with typically herbaceous vegetation, but with occasional woody plants.

MARSH, BRACKISH   Marsh flooded regularly or irregularly by water of low salt content.

MARSH, FRESHWATER   Marsh saturated or flooded by fresh water.

MARSH, HIGH   Marsh flooded irregularly by spring tides or storm surges.

MARSH, LOW   Salt marsh flooded daily by tides.

MARSH, SALT-WATER   Marsh subject to periodic flooding by salt or brackish tidal waters.

MONOCOT   A major group of flowering plants generally characterized by parallel-veined leaves, flower parts 3 or multiples thereof, and vascular bundles of stem scattered, either generally or within a band.

MORAINE   Glacial deposit of earth and stone.

MUCRONATE   Terminating abruptly in a sharp, small point or spine.

MULTIPLE FRUIT   A fruit formed from several to many flowers compactly arranged into a single structure on a common axis, as in mulberry and sweetgum.

NACL   Sodium chloride—common table salt.

NERVE   A vein. May be present in bracts, glumes, lemmas, and scales, as well as leaves.

NODE   That narrow region on the stem where a leaf or leaves attach.

NUT   An indehiscent one-seeded fruit having a hard outer wall.

NUTLET   A small nut, such as in mints and verbenas, loosely distinguished by its size; scarcely separated from an achene except by the comparative thickness of the covering.

OB-   A prefix signifying an inversion, such as obcordate, the opposite of cordate.

OBLANCEOLATE   Opposite of lanceolate, the terminal half the broader.

OBLIQUE   Having a slanting or sloping direction.

OBLONG   Elongate in form with sides parallel or nearly so, the ends more or less blunted and not tapering; wider than linear.

OBOVATE   Opposite of ovate, the terminal half the broader.

OBTUSE   A blunt or rounded point, the angle of the point being greater than 90 degrees. Compare with acute.

ORBICULAR   Circular in outline.

OVAL   Broadly elliptic.

OVARY   The part of the pistil containing the ovules, the future seeds. Also called ovulary.

OVATE   Having the outline of an egg with the broader half below the middle.

OVERWASH   Process of island flooding during storms in which sea water flows over an island to the backside.

OVOID   A 3-dimensional structure having the outline of an egg with the broader half below the middle.

OVULE   The egg-containing structure in the ovary that, after fertilization, develops into a seed.

PALMATE   Radiately arranged, ribbed, or lobed, as fingers of a hand.

PAN   See FLAT, SALT.

PANICLE   An irregularly compound raceme; e.g., sea-oats, beachgrass, *Yucca.*

PAPILIONACEOUS   Said of a corolla having a standard, wings, and keel, as in many members of Fabaceae.

PAPPUS   An outgrowth of hairs, scales, or bristles from the summit of achenes of many species of the Asteraceae. Generally considered to represent the calyx.

PARIETAL PLACENTA   See PLACENTA.

PEDATE   Palmately lobed or divided, with the lateral lobes also divided or cleft.

PEDICEL   The stalk of a single flower.

PEDUNCLE   The main flower stalk of the inflorescence supporting either a cluster of flowers or the only flower of a single-flowered inflorescence, such as in the tulip.

PERENNIAL   Plant with a life span longer than 2 growing seasons.

PERFECT   Flowers having both functional stamens and pistils.

PERFOLIATE   Descriptive of a leaf blade that completely surrounds the stem, which appears to pass through the leaf, or of 2 opposite leaves with bases fused to each other.

PERIANTH   The calyx and corolla collectively, or the calyx alone if the corolla is absent.

PETAL   One of the parts of the corolla, which may be separate or united to another petal.

PETIOLE   The attaching stalk of a leaf; sometimes absent.

PINNATE   Lobes or blades of a leaf arranged along the sides of a common axis, as the pinnae of a feather.

PISTIL   The female organ of a flower bearing one or more ovules and composed of ovary, style(s) (not always present), and stigma(s); consists of a single carpel or of 2 or more fused carpels.

PISTILLATE   Having pistils and no functional stamens.

PLACENTA   The point or line of attachment of ovules within an ovary or of seeds within a fruit. Placenta types are as follows: *Axile*—attachment is at or along the central axis of an ovary with 2 or more cells. *Basal*—one or a few ovules or seeds are attached at the base of the ovary. *Central*—attachment is to a central column in a one-celled ovary, fastened at its base but otherwise free of the ovary wall. *Parietal*—attachment is to the inside of the exterior wall of the ovary.

PLACENTATION   The arrangement of ovules within the ovary.

PLEISTOCENE   Earlier of 2 Epochs of the Quaternary Period of geologic time that began 1.5 million years ago. See HOLOCENE.

POD   A dry, dehiscent fruit.

POME   A fleshy fruit, as in crabapple or chokeberry, having several seed chambers formed from the ovary wall, part of which is parchment-like or bony in texture; the fleshy portion is formed largely from the receptacle.

PRAIRIE   A natural grassland.

PRICKLE   A small, sharp, spinelike projection that is a part of the bark or epidermis.

PRISMATIC   Like a prism, angular with flat sides.

PROSTRATE   Lying parallel to the ground.

RACEME   An inflorescence in which stalked flowers are arranged singly along a common elongated axis.

RACHIS   The main axis of a spike or of a pinnately compound leaf, excluding the petiole.

RAY   See LIGULE.

RECEPTACLE   The summit of the pedicel to which the flower parts are attached. Also the enlarged summit of the peduncle of a head to which the flowers are attached, such as in composites.

REFLEXED   Abruptly turned or bent toward the base. See RETRORSE.

REGULAR   Flowers in which the members of each set of parts (e.g., petals) are the same size and shape.

RENIFORM   Like the outline of a kidney, rounded but with a wide basal notch.

RETRORSE   Directed backward or downward. See REFLEXED.

REVOLUTE   Having the margins rolled inward toward the lower leaf surface.

RHIZOME   A horizontal underground stem; distinguished from a root by presence of nodes, buds, or scalelike leaves that are sometimes quite small.

RHOMBIC   With the outline of an equilateral parallelogram, having oblique angles.

ROOTSTOCK   Rootlike stem or rhizome, either on or in the ground, erect or horizontal.

SAGITTATE   Shaped like an arrowhead. Also see HASTATE.

SAVANNA   An area containing scattered trees and drought-resistant undergrowth.

SCABROUS   Rough or harsh to the touch due to minute stiff hairs or other projections.

SCALE   Applied to many kinds of small, thin, flat, usually dry, appressed leaves or bracts, often vestigial. Sometimes epidermal outgrowths, if disclike or flattened. Scales subtend flowers in the sedges.

SCHIZOCARP   A dry fruit that splits into 2 one-seeded halves, as in most members of the parsley family.

SCURFY   Surface with small scale-like or branlike particles.

SEPAL   One of the parts of the calyx or outer set of flower parts; may be separate or united to another sepal.

SERRATE   Having sharp sawlike teeth pointed upward or forward.

SESSILE   Without any kind of stalk.

SHEATH   A tubular structure surrounding an organ or part, such as the lower part of the leaf of grasses.

SHINGLE   See BEACH.

SHOALS   Mounds of sediment in shallow water on the ocean floor that do not emerge above water level at low tide.

SHORE   Zone from lower margin of low tide to base of dunes, including beach and berm.

SHORELINE   Point where land and water meet.

SINUATE   With a strongly wavy margin.

SLACK   See SWALE.

SLOUGH   Area of slow-moving water between older dunes.

SPATULATE   Like a spatula; somewhat broadened toward a rounded end.

SPIKE   A type of inflorescence in which stalkless flowers are attached along the sides of an elongated common axis.

SPIKELET   A small inflorescence with flowers sessile in axils of bract(s) or scale(s), which are closely arranged in 2 rows or in spirals on a common axis; a diminutive spike. The basic inflorescence of grasses and sedges.

SPIT   See BARRIER SPIT.

SPRING TIDE   Tide of maximum rise occurring near times of new and full moon.

SPUR   A saclike or tubular extension of some part of the flower.

SSP.   Subspecies.

STAMEN   The pollen-producing organ of a flower, usually consisting of anther and filament.

STAMINATE   Having stamens and no functional pistil.

STANDARD   The upper dilated petal of a papilionaceous corolla.

STELLATE   Star-shaped. Applicable where several similar parts spread out from a common center. Usually applied to branched hairs.

STIGMA   The pollen-receptive part of a pistil, often enlarged, usually at the tip of the style.

STIPE   Stalk portion of a pistil or fruit between the calyx and corolla and the seed-bearing part of the ovary.

STIPEL   Stipule of a leaflet of a compound leaf.

STIPULES   A pair of structures, usually small, on the base of the petiole of a leaf or on the stem near the petiole, or on both; sometimes fused together.

STOLON   A slender stem that runs along the ground surface, or just below it and produces a new plant at its tip.

STOMATE   Minute opening in epidermis of plant part, usually the leaf, through which gas exchange takes place.

STONE   A seed and its bony covering, as in a peach or cherry.

STYLE   That portion of the pistil between the stigma and ovary, often elongate and threadlike, sometimes apparently absent.

SUBULATE   Awl-shaped; narrowly triangular, tapering from base to apex.

SUCCULENT   Soft, juicy, fleshy, and thickened.

SUPERIOR   Said of an ovary that stands free of other floral organs.

SWALE   Small depressed area behind dune (usually primary dune) where ground water approaches the surface. After heavy rains these depressions may become pools.

SWAMP   Low, wooded, freshwater wetland with saturated soil or standing water.

SYNONYM   A taxonomic name that has been superseded by another name.

TECTONIC   Pertaining to geologic structure involving folding, faulting, and deformation of earth's crust.

TENDRIL   A threadlike structure, usually in a climbing plant, that may twine around an object and support the plant.

TERETE   Circular in cross section.

THORN   A hardened, pointed branch.

TRUNCATE   An apex or base nearly or quite straight across, as if cut off.

TUBERCULATE   Bearing small raised places.

TWO-RANKED   Alternate or opposite leaves attached on a stem in 2 opposite vertical rows, thus the places of attachment lying in one plane.

UMBEL   A type of flower cluster in which flower stalks of approximately equal length arise from the same level on the stem like ribs of an umbrella. Characteristic of the parsley family. A compound umbel has a second group of umbels set on the first and sometimes a third group on the second.

UNISEXUAL   Of one sex, with either stamens or pistils only.

VALVATE   Meeting at edges without overlapping, as in some sepals.

VASCULAR   Pertaining to vessels or ducts of the conducting system within plants.

VISCID   Having a sticky, glutinous consistency.

WATER TABLE   Transition line between aerated and water-saturated soil.

WHORL   Three or more structures (leaves, stems, etc.) in a circle, not spiralled.

WING   A thin membranous expansion or flat extension of an organ; lateral petal in Fabaceae and Polygalaceae.

XEROPHYTE   Plant adapted to an arid environment.

Keys

The keys are designed to identify the family or genus to which the species belongs, and sometimes the species itself. Identification is possible by making a series of choices between a set of two or more contrasting characteristics numbered identically. These sets of numbers with their characteristics are called couplets. Following each half of the couplet, a family, genus, or species is given; or a number indicating the next couplet to be used. *Page location of the family , genus, or species is shown in the index at the back of the book.* Family names are also displayed at the top of pages throughout.

    The keys can be likened to driving on a road system having a series of Y connections and no crossroads. Each Y connection requires a choice that should be made with care, otherwise it may be necessary to turn back and recheck decisions. In the case of the keys, this is facilitated by a number in parentheses placed after the first of the two identical numbers of a couplet. For example, in Key I for the dicots we find couplet 5 written as 5(4) and 5. This indicates that the last choice made in reaching couplet 5 was couplet 4. Similarly, the last choice before arriving at couplet 15 [presented as 15(3) and 15] was couplet 3. By using the numbers in parentheses, one may return easily to previous decisions and try a different route. Considerable practice may be necessary to become proficient with the keys but the time required will be amply repaid.

Key I Dicots with Separate Petals

1.	Petals numerous	Nymphaeaceae
1.	Petals 4–5 or absent	2
2(1).	Petals of equal size and shape, or absent	3
2.	Petals of unequal size or shape, or both	27
3(2).	Fruit 1-celled; 1 to many per flower	4
3.	Fruit 2- to many-celled; 1 per flower	15
4(3).	Fruits with 1 seed	5
4.	Fruits with 2 to many seeds	9

5(4).   Stipules forming sheath around stem just above
        nodes . . . .   Polygonaceae
5.   Stipules absent, or present and not forming sheath 6
6(5).   Fruits 1 per flower 7
6.   Fruits 2 to many per flower 8
7(6).   Perianth fleshy . . . .   Chenopodiaceae
7.   Perianth thin, membranous . . . .   Amaranthaceae
8(6).   Stipules present, hypanthium present . . . .   Rosaceae
8.   Stipules absent, hypanthium absent . . . .   *Ranunculus*
9(4).   Leaves thick, succulent . . . .   Crassulaceae
9.   Leaves flattened, not succulent 10
10(9). Carpel 1, fruit a legume . . . .   Fabaceae
10.   Carpels 2–5, fruit a capsule 11
11(10). Fruit a capsule with a lid opening along an even
        circular line . . . .   *Portulaca*
11.   Fruit a capsule opening along longitudinal lines 12
12(11). Placenta central . . . .   Caryophyllaceae
12.   Placenta axile or parietal 13
13(12). Styles 3–5, sometimes united near base . . . .   Hypericaceae
13.   Style 1 14
14(13). Carpels 3 . . . .   Cistaceae
14.   Carpels 2 . . . .   *Corydalis*
15(3). Flowers and fruits in umbels 16
15.   Flowers and fruits not in umbels 17
16(15). Fruits dry, with two 1-seeded sections that finally
        split apart . . . .   Apiaceae
16.   Fruit fleshy, a berrylike drupe . . . .   *Aralia*
17(15). Fruit fleshy, a berry with parietal placenta . . . . .   *Passiflora*
17.   Fruit dry 18
18(17). Carpels 3, each carpel of fruit 1-seeded . . . . .   Euphorbiaceae
18.   Carpels 2, 4 or more 19
19(18). Carpels 2 20
19.   Carpels 4 or more 21
20(19). Placenta parietal . . . .   Brassicaceae
20.   Placenta axile . . . .   Lythraceae
21(19). Plants without chlorophyll, color almost white to almost red, or
        yellow to lavender . . . .   *Monotropa*
21.   Plants with chlorophyll 22
22(21). Stamens mostly united, forming a cylinder around ovary and
        style . . . .   Malvaceae
22.   Stamens not forming such a cylinder 23
23(22). Hypanthium present 24
23.   Hypanthium absent 25
24(23). Hypanthium urn-shaped . . . .   *Rhexia*
24.   Hypanthium not urn-shaped . . . .   Onagraceae
25(23). Leaves simple, narrow, entire . . . .   *Linum*

25.   Leaves compound, or simple and deeply lobed   26
26(25).   Leaves deeply lobed   . . . .   *Geranium*
26.   Leaves with 3 leaflets   . . . .   *Oxalis*
27(2).   Carpel 1, seeds in one row, fruit a legume   . . . . .   Fabaceae
27.   Carpels 2 or 3, fruit not a legume   28
28(27).   Ovary and fruit 1-celled, seeds few to many   . . . . .   *Viola*
28.   Ovary and fruit 2-celled; seeds 2, one in each cell   . .   *Polygala*

Key II   Dicots with United Petals

1.   Flowers in heads surrounded by bracts, ovaries inferior, fruits 1-
          seeded   . . . .   Asteraceae
1.   Flowers not in heads, or if in heads the ovaries superior or the
          fruits more than 1-seeded   2
2(1).   Ovary superior   3
2.   Ovary inferior   20
3(2).   Corolla lobes equal in size and shape   4
3.   Corolla lobes unequal in size or shape, or both   17
4(3).   Leaf blades floating, nearly circular . . . . . .   *Nymphoides*
4.   Leaf blades not floating   5
5(4).   Plants leafless, almost white to orange, parasitic . . .   *Cuscuta*
5.   Plants with leaves, not parasitic   6
6(5).   Calyx white, membranous; fruit with 1 seed   . . . .   *Limonium*
6.   Calyx not as above; fruit with 2 or more seeds, or if 1-seeded, the
          calyx not white   7
7(6).   Flowers in umbels; fruit a follicle . . . . . .   Asclepiadaceae
7.   Flowers not in umbels; fruit usually not a follicle   8
8(7).   Fruit of 1–4 nutlets   9
8.   Fruit a capsule, follicle, or berry   10
9(8).   Leaves opposite, stamens 4 . . . .   Verbenaceae
9.   Leaves alternate, stamens 5 . . . .   *Heliotropium*
10(8).   Leaves alternate or basal   11
10.   Leaves opposite or whorled   13
11(10).   Leaves basal, flowers in spikes, fruit a capsule opening along an
          even circular line . . . .   *Plantago*
11.   Leaves mostly not basal; flowers not in spikes; fruit a berry, or
          capsule opening along longitudinal lines   12
12(11).   Plants climbing or trailing; fruit with 2–6 seeds,
          1–2 per carpel . . . .   Convolvulaceae
12.   Plants erect; fruit with many seeds, several to many
          per carpel   . . . .   Solanaceae
13(10).   Ovary and fruit 1-celled   14
13.   Ovary and fruit with 2 or more cells   16
14(13).   Placenta central   . . . .   Primulaceae
14.   Placenta parietal   15

15(14). Placenta 1, fruit splitting on 1 longitudinal line   . . . *Apocynum*
15.   Placentas 2, fruit splitting on 2 longitudinal lines . . . . *Sabatia*
16(13). Corolla lobes 5, stamens 5   . . . . Primulaceae
16.   Corolla lobes 4, stamens 4   . . . . *Polypremum*
17(3). Plants leafless above substrate, small bracts present, corollas
       yellow to deep pink to purple . . . . *Utricularia*
17.   Plants with leaves above substrate, corolla of various colors .   18
18(17). Fruit a many-seeded capsule   . . . . Scrophulariaceae
18.   Fruit of 1–4 nutlets or of 2 dry, fused, 1-seeded sections that
       finally split apart   19
19(18). Flowers sessile, in elongated terminal spikes or in terminal heads
       under 9 mm diameter . . . . Verbenaceae
19.   Flowers pedicellate; or sessile or nearly so in tight axillary
       clusters, or in heads over 10 mm wide . . . . . . Lamiaceae
20(2). Leaves opposite . . . . Rubiaceae
20.   Leaves alternate   21
21(20). Plant a vine with tendrils . . . . *Melothria*
21.   Plant erect, without tendrils   . . . . Campanulaceae

Key III   Herbaceous Monocots with Sepals and Petals

1.   Flowers and fruits in terminal umbels, ovary superior 2
1.   Flowers and fruits not in umbels; if umbel-like, the ovary
       inferior   3
2(1). Ovules and seeds 2 in each carpel, plants with strong onion
       odor   . . . . *Allium*
2.   Ovules and seeds several in each carpel, faint onion odor may be
       present when plant is fresh . . . . *Nothoscordum*
3(1). Flowers and fruits in terminal heads, 1 per stem; leaves basal . 4
3.   Flowers and fruits not in heads, or if in headlike clusters then
       more than one head per stem; leaves usually not
       entirely basal   5
4(3). Flower heads buttonlike, whitish to grayish; flowers very small,
       numerous, surrounded by bracts at base . . . . . . *Eriocaulon*
4.   Flower heads conelike, of numerous tightly overlapping scalelike
       bracts, from each of which protrudes a single yellow flower,
       usually only one appearing at a time . . . . *Xyris*
5(3). Stamens many, sometimes only stamens in some flowers and only
       pistils in other flowers . . . . *Sagittaria*
5.   Stamens 6 or less   6
6(5). Petals of equal size and shape 7
6.   Petals of unequal size or shape, or both   16
7(6). Plants epiphytic, bearing grayish peltate scales   . . . *Tillandsia*
7.   Plants not epiphytic, not bearing such scales 8
8(7). Ovary superior   9

8.   Ovary inferior 14
9(8).   Flowers and fruits in spikelike racemes . . . . *Triglochin*
9.   Flowers not in spikelike racemes 10
10(9).   Sepals green; petals blue to purple or purplish-pink,
         rarely white . . . . *Tradescantia*
10.   Sepals and petals alike in color, the colors various 11
11(10). Sepals and petals chaffy or scalelike, greenish-brown to dark
         reddish-brown . . . . *Juncus*
11.   Sepals and petals not chaffy or scalelike, color otherwise than
         above 12
12(11). At least some leaves whorled; perianth orange to orange-red, 5–9
         cm long . . . . *Lilium*
12.   All leaves alternate, perianth white to cream-colored . . . 13
13(12). Leaves 2 or 3 . . . . *Maianthemum*
13.   Leaves several to many . . . . *Smilicina*
14(8).   Stamens 6 . . . . *Hypoxis*
14.   Stamens 3 15
15(14). Stems winged; perianth segments obviously 6, stamens clearly
         evident . . . . *Sisyrinchium*
15.   Stems not winged; perianth segments appearing to be 9, the 3
         stigmas petal-like and hiding the stamens . . . . *Iris*
16(6).   Flowers arising from a closely folded bract; 2 petals blue to light
         purple, the third one smaller and nearly colorless . *Commelina*
16.   Flowers not arising from a closely folded bract, all petals with
         definite color 17
17(16). Most of flower a definite yellow, about 7 cm long; plants usually
         about 1 m tall . . . . *Canna*
17.   General color of flowers not a definite yellow, plants usually
         under 1 m tall 18
18(17). Stamens with anthers 1–2, not readily evident, being united with
         pistil and forming a column . . . . *Orchidaceae*
18.   Stamens with anthers 6, 3 long and evident, 3 short and
         sometimes not conspicuous 19
19(18). Leaf blades suborbicular to broadly elliptic, petiole usually
         inflated conspicuously . . . . *Eichhornia*
19.   Leaf blades cordate-ovate to lanceolate, petiole
         not inflated . . . . *Pontederia*

Key IV Herbaceous Monocots without Sepals and Petals, or sepals and petals small and of a form different from that ordinarily recognized

1. Flowers very small, numerous, arranged at top of stem in coarse cylindrical clusters . . . . *Typha*

1. Flowers usually small but not numerous, in coarse cylindrical clusters unless accompanied by conspicuous bristles 2

2(1). Poaceae, the grasses. Leaves 2-ranked; internodes hollow, rarely solid (see p. 323 for other differences) 3

2. Cyperaceae, the sedges. Leaves 3-ranked; internodes solid, rarely hollow 35

3(2). Plants to 2.5 m tall; spikelets in spikes or spikelike racemes, unisexual, the sexes not intermixed, the female spike breaking up into beadlike units . . . . *Tripsacum*

3. Plants tall or short; if spikelets in spikes or spikelike racemes the spikelets bisexual and alike 4

4(3). Spikelets closely arranged in 1-sided spikes or spikelike racemes 5

4. Spikelets in racemes, panicles, or 2-sided spikes 9

5(4). Spike solitary at end of stem or of lateral branch . *Stenotaphrum*

5. Spikes 2 or more on upper part of stem or lateral branch . . . 6

6(5). Spikelets consisting of an inner unit that is conspicuously firm, shiny, veinless and rounded; inner unit closely clasped by 2 veined membranous "bracts" . . . . *Paspalum*

6. Spikelets lacking inner unit described above 7

7(6). Spikes arranged in racemes . . . . *Spartina*

7. Spikes diverging from a common point; sometimes 1–2 spikes below this group 8

8(7). Spikelets composed of several similar greenish units . *Eleusine*

8. Spikelets with few units, one a conspicuous dark-brown color . . . . *Eustachys*

9(4). Spikelets in hard burs bearing sharp, stiff spines . . . *Cenchrus*

9. Spikelets not in such burs 10

10(9). Spikelets in cylindrical clusters containing conspicuous bristles or awns 11

10. Spikelets not in cylindrical clusters, sometimes in nearly cylindrical panicles but bristles and awns lacking 12

11(10). Bristles below base of spikelets . . . . *Setaria*

11. Bristles absent, awns on end of units of spikelets . . . . *Elymus*

12(10). Spikelets with an inner unit that is conspicuously firm, shiny, veinless and rounded; this unit closely clasped by 3 veined membranous "bracts," the lowest one often small 13

12. Spikelets lacking inner unit described above 15

13(12). Spikelets in slender panicles, second glume bulging at base, appearing much like baggy knees . . . . *Sacciolepis*

13. Spikelets in broader panicles, the glume little or not bulging . 14

14(13). No awns on spikelets . . . . *Panicum*
14. Spikelets with at least second glume awned . . . . *Echinochloa*
15(12). Plants over 2 m tall, with large terminal panicles 16
15. Plants under 2 m tall, panicles usually not large 20
16(15). Spikelets with long fine hairs arising from base of spikelet or from inside it 17
16. Spikelets without long fine hairs; a single awn may be present 18
17(16). Long fine hairs arising from base of spikelets . . . . *Erianthus*
17. Long fine hairs arising from inside spikelet . . . . *Phragmites*
18(16). Spikelets in narrow panicles . . . . *Sorghastrum*
18. Spikelets in broad panicles 19
19(18). Spikelets unisexual, the female on upper branches, male on lower branches and falling soon after pollen is shed . . . . . *Zizania*
19. Spikelets unisexual, the female and male on same branches, the male falling early . . . . *Zizaniopsis*
20(15). Spikelets in pairs in racemes; one spikelet sessile, the other pedicelled and smaller to absent, in which case represented by its pedicel . . . . *Andropogon, Schizachyrium*
20. Spikelets not in pairs, in panicles 21
21(20). Spikelets with awns, awn on lemma may be hidden by glumes 22
21. Spikelets awnless 27
22(21). Glumes, exclusive of any awns, longer than lemmas, the upper of the 2 lemmas with a short curved or hooked awn . . . *Holcus*
22. Glumes longer or shorter than lemmas, awns not strongly curved or hooked 23
23(22). Plants usually over 80 cm tall, awn from inside firm shiny glumes . . . . *Sorghum*
23. Plants usually under 80 cm tall, awns and glumes not as above 24
24(23). Spikelets 1-flowered 25
24. Spikelets 2–13-flowered 26
25(24). Panicle compact . . . . *Polypogon*
25. Panicle quite open . . . . *Muhlenbergia*
26(24). Plants erect, lemmas 5–13, awn on tip of lemma . . . . *Vulpia*
26. Plants usually spreading, lemmas 2–4, awn from back of notched tip on lemma . . . . *Triplasis*
27(21). Lemmas 1 per spikelet 28
27. Lemmas 2 or more per spikelet 30
28(27). Spikelets 10–15 mm long . . . . *Ammophila*
28. Spikelets 2–3 mm long 29
29(28). Glumes longer than lemma, panicle open . . . . . *Agrostis*
29. Glumes shorter than lemma; panicle compact, often spikelike . . . . *Sporobolus*
30(27). Lemmas as broad as long, horizontally spreading . . . *Briza*

30.   Lemmas longer than broad, ascending   31

31(30). Spikelets conspicuously flattened, about 10 mm wide, in broad open panicles . . . . *Uniola*

31.   Spikelets not conspicuously flattened, or if so less than 8 mm wide and panicle compact or elongated or both   32

32(31). Spikelets 3.5–8 mm wide   33

32.   Spikelets 1.5–3 mm wide   34

33(32). Panicle erect, plants of open salt or brackish habitats . *Distichlis*

33.   Panicle arching, plants of wooded habitats . . *Chasmanthium*

34(32). Perennials, lemmas 3-veined . . . . *Eragrostis*

34.   Annuals, lemmas 5-veined . . . . *Poa*

35(2).   Cyperaceae. Top of plants bearing 4–10 prominently white-based leaves (bracts) that surround the tight cluster of spikelets . . . . *Dichromena*

35.   Plants lacking such white-based bracts   36

36(35). Plant with a single spikelet at very top of an apparently leafless stem, the leaves consisting of sheaths only . . . . *Eleocharis*

36.   Plant with more than one spikelet at upper part; or if only one, the stem appearing to be prolonged above it   37

37(36). Mature fruits bony and white (or nearly so) achenes . . *Scleria*

37.   Mature fruits not white   38

38(37). Plants 1–3 m tall, spikelets in conspicuous clusters arising from axils of the well-separated leaves of the upper part of the stem, leaf margins with small dangerous saw-teeth . . . *Cladium*

38.   Plants usually under 1 m tall, occasionally to 2 m tall; spikelets usually in terminal clusters; if plants over 1 m tall, the leaf margins at most minutely sharply serrate   39

39(38). Flowers in spikes, unisexual, the sexes in different spikes or different parts of the same spike; a saclike structure surrounding the pistil (the 2–3 stigmas protruding) and later the achene . . . . *Carex*

39.   Flowers bisexual, no saclike structure around pistil or achene   40

40(39). Spikelets with 2-ranked scales . . . . *Cyperus*

40.   Spikelet scales spirally arranged   41

41(40). Spikelets composed of few scales, achene surmounted by a beak . . . . *Rhynchospora*

41.   Spikelet composed of many scales, achene not topped by a beak   42

42(41). Behind each of the spiralled scales of the spikelet are 3 scales with expanded blades on stalks that arise under the achene . . *Fuirena*

42.   Stalked scales absent from around achene   43

43(42). Bristles arising from under achene . . . . *Scirpus*

43.   No bristles arising from under achene; filaments that are sometimes similar may be found there . . . . *Fimbristylis*

Key V   Groups of Woody Plants

1.   Leaves under 4.5 mm wide; simple; needlelike, awl-like, or scalelike . . . . Group A
1.   Leaves over 4.5 mm wide, simple or compound, blades linear to orbicular 2
2(1).   Leaves with all veins parallel . . . . Group B
2.   Leaves with netted veins. In some species each leaf has 2 sets of parallel major veins 3
3(2).   Woody vines . . . . Group C
3.   Shrubs or trees 4
4(3).   Leaves compound . . . . Group D
4.   Leaves simple 5
5(4).   Leaves opposite or whorled . . . . Group E
5.   Leaves alternate 6
6(5).   Leaves with 3–7 large veins arising from one place at or near base . . . . Group F
6.   Leaves with 1 large vein at base of blade 7
7(6).   Leaves evergreen . . . . Group G
7.   Leaves deciduous, or dead and persisting 8
8(7).   Leaves and leaf scars on vigorous twigs 2-ranked . . . Group H
8.   Leaves and leaf scars on vigorous twigs not 2-ranked 9
9(8).   Leaf margins entire . . . . Group I
9.   Leaf margins serrate (sometimes minutely) to dentate . Group J

Group A: Leaves under 4.5 mm wide, needlelike, awl-like, scalelike, or linear to elliptic

1.   Leaves needlelike, in tight clusters of 3–5, surrounded by a sheath of scale leaves . . . . *Pinus*
1.   Leaves scalelike, awl-like, or linear and not in tight clusters surrounded by a sheath 2
2(1).   Leaves opposite or whorled 3
2.   Leaves alternate 6
3(2).   Leaves awl-like or scalelike 4
3.   Leaves linear to elliptic or oblong 5
4(3).   Stems of current year succulent . . . . *Sarcocornia*
4.   Stems all woody . . . . *Chamaecyparis, Juniperus*
5(3).   Leaves opposite . . . . *Hypericum*
5.   Leaves whorled . . . . *Ceratiola, Corema, Kalmia*
6(2).   Stems fleshy, jointed . . . . *Opuntia*
6.   Stems not fleshy 7
7(6).   Leaves succulent . . . . *Lycium*
7.   Leaves not succulent 8
8(7).   Plants low, spreading . . . . . *Empetrum, Hudsonia, Vaccinium*

8.	Plants erect 9
9(8).	Leaves under 2 mm long . . . . *Tamarix*
9.	Leaves over 15 mm long . . . . *Baccharis angustifolia*

Group B: Leaves over 4.5 mm wide, veins parallel

1.	Leaves lance-shaped, sessile . . . . *Yucca*
1.	Leaves fan-shaped, petioled 2
2(1).	Petioles armed with sharp spines . . . . *Serenoa*
2.	Petioles not armed . . . . *Sabal*

Group C: Woody vines

1.	Leaves simple 2
1.	Leaves compound 4
2(1).	Stems not twining . . . . *Smilax, Vitis*
2.	Stems twining 3
3(2).	Leaves opposite . . . . *Gelsemium, Lonicera*
3.	Leaves alternate . . . . *Berchemia*
4(1).	Leaves opposite 5
4.	Leaves alternate 6
5(4).	Leaflets with blades 2 . . . . *Bignonia*
5.	Leaflets 7–13 . . . . *Campsis*
6(4).	Leaflets 3 (POISONOUS TO TOUCH) . . . . . *Toxicodendron*
6.	Leaflets many . . . . *Ampelopsis, Parthenocissus*

Group D: Shrubs and trees with compound leaves

1.	Leaves opposite 2
1.	Leaves alternate 3
2(1).	Leaves palmately compound . . . . *Aesculus*
2.	Leaves pinnately compound . . . . *Sambucus*
3(1).	Leaves twice compound 4
3.	Leaves once compound 5
4(3).	Leaflets many, under 5 mm wide . . . . *Acacia*
4.	Leaflets many, over 10 mm wide . . . . *Aralia*
5(3).	Stems with prickles 6
5.	Stems without prickles 7
6(5).	Stems biennial, fruit juicy . . . . *Rubus*
6.	Stems perennial, fruit not juicy . . . . *Rosa*
7(5).	Leaflets serrate 8
7.	Leaflets entire, sometimes lobed, or with large teeth 9
8(7).	Fruit a nut inside a 4-sectioned husk . . . . *Carya*

8.	Fruit a small drupe	*Rhus*
9(7).	Leaflets 3 (POISONOUS TO TOUCH)	*Toxicodendron*
9.	Leaflets 7 or more	10
10(9).	Fruit a small drupe	11
10.	Fruit a legume	12
11(10).	Stems glabrous (POISONOUS TO TOUCH) . . .	*Toxicodendron*
11.	Stems densely fine hairy	*Rhus*
12(10).	Fruit conspicuously 4-winged	*Daubentonia*
12.	Fruit not 4-winged	13
13(10).	Fruit under 1 cm long	*Amorpha*
13.	Fruit over 4 cm long	*Robinia*

Group E: Leaves simple, over 4.5 mm wide, opposite

1.	Leaves palmately veined	*Acer*
1.	Leaves pinnately veined	2
2(1).	Flowers and fruits in heads	3
2.	Flowers and fruits not in heads	8
3(2).	Mature fruits dry	4
3.	Mature fruits fleshy	6
4(3).	Fruits in dense globose heads	*Cephalanthus*
4.	Fruits not in globose heads	5
5(4).	Disc flower bracts noxiously spiny-tipped	*Borrichia*
5.	No spiny-tipped bracts	*Iva*
6(3).	Leaves thick, fleshy	*Batis*
6.	Leaves thin	7
7(6).	Stem with scattered prickles	*Lantana*
7.	Stem lacking prickles	*Cornus*
8(2).	Partial parasite	*Phoradendron*
8.	Not a parasite	9
9(8).	Ovary inferior	10
9.	Ovary superior	11
10(9).	Petals separate	*Cornus*
10.	Petals united	*Viburnum*
11(9).	Leaves 3-veined near base	*Sageretia*
11.	Leaves 1-veined near base	12
12(11).	Leaves serrate	*Callicarpa*
12.	Leaves entire	13
13(12).	Leaves to 5 cm long	*Forestiera*
13.	Leaves over 7 cm long	*Osmanthus*

Group F: Leaves alternate, palmately veined

1.	Shrubs with straight prickles, at least at base	*Ribes*
1.	Trees; if shrubs, lacking prickles	2
2(1).	Leaves with 4–7 prominent large veins arising from one place at base of blade	3
2.	Leaves with 3 such veins	5
3(2).	Petioles swollen at both ends, blades entire and heart-shaped	*Cercis*
3.	Petioles not swollen at ends	4
4(3).	Blades of leaves 5-lobed, the lobes acute	*Liquidambar*
4.	Blades of leaves not lobed	*Populus*
5(2).	Stems with thorns (may be quite scattered)	*Crataegus*
5.	Stems lacking thorns	6
6(5).	Pith of mature twigs with closely and usually irregularly spaced chambers separated by soft partitions	*Celtis*
6.	Pith of mature twigs continuous	7
7(6).	Margins of leaf blades doubly serrate	*Betula*
7.	Margins simply serrated or crenate	8
8(7).	Foliage, buds, and twigs spicy-aromatic	9
8.	Foliage, buds, and twigs not spicy-aromatic	10
9(8).	Leaves with yellowish calluslike growths in principal vein angles on upper side of leaf	*Cinnamomum*
9.	Leaves lacking such calluslike growths	*Sassafras*
10(8).	True terminal bud lacking, the bud at end of stem an axillary bud	*Morus*
10.	Terminal bud present	*Populus*

Group G: Leaves simple, alternate, evergreen

1.	Plants with creeping stems	2
1.	No creeping stems	5
2(1).	Sepals persisting at base of fruit	3
2.	Sepals present near summit of fruit	4
3(2).	Corolla 4-lobed, fruit a capsule	*Epigaea*
3.	Corolla 5-lobed, fruit fleshy	*Arctostaphylos*
4(2).	Leaves 2–6 cm long, corolla 5-lobed	*Gaultheria*
4.	Leaves to 1.8 cm long, corolla 4-lobed . .	*Vaccinium macrocarpon*
5(1).	Plants under 50 cm tall, or stems slender and under 1.5 m tall .	6
5.	Plants over 50 cm tall	8
6(5).	Many grayish peltate scales on leaf underside. . . .	*Cassandra*
6.	Lower leaf surface shiny, lacking scales	7
7(6).	Leaf apex rounded or shallowly notched	*Licania*
7.	Leaf apex pointed . . . . . *Vaccinium myrsinites* and *V. darrowii*
8(5).	Leaves succulent	*Iva imbricata*

8.   Leaves not succulent 9
9(8).   Stipules and stipule scars encircling twig . . . . *Magnolia*
9.   Stipules absent or, if present, not encircling twig 10
10(9).   Foliage or twigs or both notably aromatic when crushed
          or broken 11
10.   Not aromatic, but may have bitter almond odor 12
11(10).   Lower leaf surface with small yellowish to brownish glandular
          dots . . . . *Myrica*
11.   Leaves without glandular dots . . . . *Persea*
12(10).   Underside of leaf blades with grayish to yellowish shieldlike
          scales . . . . *Lyonia ferruginea* and *L. fruticosa*
12.   Underside of leaf blades lacking scales 13
13(12).   Twigs sharply angled . . . . *Lyonia lucida*
13.   Twigs not sharply angled 14
14(13).   Pith of twigs chambered . . . . *Symplocos*
14.   Pith of twigs continuous 15
15(14).   Fresh twigs with rank bitter-almond odor and taste when
          broken . . . . *Prunus caroliniana*
15.   Fresh twigs without this taste or odor 16
16(15).   Stipules and stipule scars present 17
16.   Stipules and stipule scars absent 18
17(16).   Vigorous twigs with one leaf at end (leaves often clustered at tip
          of short twigs) . . . . *Ilex*
17.   Vigorous twigs with 2–4 leaves clustered at tip . . . *Quercus*
18(16).   Largest leaves with a few to several conspicuous
          teeth . . . . *Baccharis*
18.   Largest leaves entire or at most inconspicuously serrate . . 19
19(18).   Twigs with thorns, sometimes quite scattered; juice of vigorous
          twigs milky . . . . *Bumelia*
19.   Twigs thornless, juice not milky 20
20(19).   Netted vein system of leaf invisible or barely distinct on upper
          surface . . . . *Cliftonia*
20.   Netted vein system of leaf blade conspicuous on
          upper surface 21
21(20).   Vigorous twigs with one leaf at end,
          fruit a dry berry . . . . *Vaccinium arboreum*
21.   Vigorous twigs with 2–4 leaves clustered at tip, fruit a small
          capsule . . . . *Cyrilla*

Group H: Leaves simple, alternate, deciduous, 2-ranked

1.   Lateral buds stalked . . . . *Hamamelis*
1.   Lateral buds sessile 2
2(1).   Leaf blades entire . . . . *Asimina*
2.   Leaf blades serrate to dentate 3

3(2).	Teeth of leaf margins of same number as lateral veins, a lateral vein ending at tip of each tooth 4
3.	Teeth of leaf margins more abundant than lateral veins . . . 5
4(3).	Stipules and stipule scars nearly meeting around twig . . *Fagus*
4.	Stipules and stipule scars much shorter . . . . . . *Castanea*
5(3).	Lenticels of vigorous 2- to 3-year-old twigs horizontally long, distinct . . . . *Betula*
5.	Lenticels of vigorous 2- to 3-year-old twigs circular, often indistinct; or apparently absent 6
6(5).	Leaf blades ovate to nearly orbicular . . . . *Corylus*
6.	Leaf blades narrower 7
7(6).	True terminal bud present . . . . *Amelanchier*
7.	True terminal bud absent, the end bud axillary 8
8(7).	Leaf blades asymmetrical at base . . . . *Ulmus*
8.	Leaf blades symmetrical at base 9
9(8).	Trunk with flaking bark, leaves with fine soft hairs scattered on underside . . . . *Ostrya*
9.	Trunk smooth and fluted with musclelike ridges, leaves glabrous below except for small tufts of hairs in vein axils . . . *Carpinus*

Group I: Leaves simple, alternate, deciduous, not 2-ranked, entire

1.	Entire plant except upper surface of leaves densely covered with star-shaped hairs . . . . *Croton*
1.	Plant glabrous or, if hairy, not as above 2
2(1).	Leaves as long as wide or nearly so, a pair of glands on upper side at juncture of petiole and blade . . . . *Sapium*
2.	Leaves longer than wide and no such glands present 3
3(2).	True terminal bud present 4
3.	True terminal bud absent, the tip bud axillary 9
4(3).	Fresh foliage, buds, and twigs spicy-aromatic when crushed or tasted . . . . *Sassafras*
4.	Fresh foliage and twigs not spicy-aromatic 5
5(4).	Fresh foliage and twigs with a bitter-almond taste and odor when crushed and broken . . . . *Prunus caroliniana*
5.	Fresh foliage and twigs lacking such a taste and odor 6
6(5).	Pith chambered . . . . *Symplocos*
6.	Pith continuous 7
7(6).	Pith continuous but diaphragmed . . . . *Nyssa*
7.	Pith not diaphragmed 8
8(7).	Bud with 2, or rarely 3, nearly valvate scales . *Cornus alternifolia*
8.	Buds with several overlapping scales . . . . *Quercus*
9(3).	Shrub with yellowish glandular dots on pedicels and underside of leaves . . . . *Gaylussacia*
9.	Shrubs or trees lacking such glandular dots 10

10(9).   Stipules and stipule scars present, these quite small   . . . *Ilex*

10.   Stipules and stipule scars absent   11

11(10).   Pith continuous but diaphragmed, fruit a berry with one to several large flat seeds . . . . *Diospyros*

11.   Pith not diaphragmed, fruit a berry with several small seeds or one large rounded seed   12

12(11).   Twigs with thorns, sometimes quite scattered; fruit a 1-seeded berry resembling a cherry . . . . *Bumelia*

12.   Twigs without thorns, fruit a several-seeded berry   . . *Vaccinium*

Group J: Leaves simple, alternate, deciduous, not 2-ranked, serrate to dentate

1.   Lateral buds evident and stalked . . . . *Alnus*

1.   Lateral buds not stalked, sometimes quite inconspicuous . . .   2

2(1).   Twigs long remaining green, about 8-ribbed; leaf blades with a few to several conspicuous teeth . . . . *Baccharis*

2.   Young twigs often green, older ones dark; if not turning dark after first year, then not 8-ribbed   3

3(2).   True terminal bud absent, the tip bud axillary   4

3.   True terminal bud present   6

4(3).   Bud scale 1, caplike; bundle scars 3; stipules present   . . . *Salix*

4.   Bud scales 2–6, overlapping; bundle scar 1, stipules absent . .   5

5(4).   Fruit a capsule, some persisting into following year . . . *Lyonia*

5.   Fruit a berry . . . . *Vaccinium*

6(3).   Bundle scar 1, U-shaped, conspicuously raised and protruding; pith about half width of year-old twigs, with a network of firm strands . . . . *Clethra*

6.   Bundle scars 1 to many but not raised, U-shaped, or protruding; pith usually quite small   7

7(6).   No stipule scars; stipules may be present but attached to petiole   8

7.   Stipules attached to twig, stipule scars therefore present . .   10

8(7).   Stipules absent, bundle scar 1, fruit 5–8 small separate follicles . . . . *Spiraea*

8.   Stipules attached to lower end of petiole   9

9(8).   Small tree, fruit a pome 25–35 mm long . . . . *Malus*

9.   Shrub, fruit a pome 6–10 mm long . . . . *Aronia*

10(7).   Fresh twigs with a bitter-almond taste and odor when crushed or broken . . . . *Prunus*

10.   Fresh twigs without this taste or odor   11

11(10).   Thorns present, often quite scattered . . . . . . *Crataegus*

11.   Thorns absent   12

12(11).   Bundle scar 1; stipules soon turning dark, persistent, very short, sharp-pointed; fruit a berrylike drupe . . . . *Ilex*

12.   Bundle scars 3 or more; stipules falling early; fruit a nut, or small capsules in dense elongate clusters 13

13(12). Leaf blades about as broad as long, apex acute to acuminate; fruit small capsules in clusters . . . . *Populus*

13.   Leaf blades not as broad as long or, if so, the apex rounded; fruit an acorn, a type of nut, borne in scaly cups . . . . . *Quercus*

Illustration numbers of the style [1] correspond to the numbers shown with the species descriptions so as to allow cross-reference between illustrations and text.

[1] *Saururus cernuus* × 1/2

[2] *Boehmeria cylindrica* × 1/3

[3] *Rumex acetosella* × 1/4

[4] *Rumex hastatulus* × 1/8

[5] *Rumex verticillatus* × 1/3

[6] *Rumex crispus* × 1/4

[7] *Rumex orbiculatus* × 1/25

[8] *Polygonum aviculare* × 1⅔

[9] *Polygonum punctatum* × 2/5

[10] *Polygonum pensylvanicum* × 1/5

[11] *Polygonum hydropiperoides* × 2/5

[12] *Polygonum sagittatum* × 2/5

[13] *Polygonum convolvulus* × 1/2

[14] *Polygonella articulata* × 3/5

[15] *Chenopodium album* × 1/5

[16] *Chenopodium ambrosioides* × 1/3

[17] *Atriplex arenaria* × 1

[18] *Atriplex patula* × 1/4

[19] *Salicornia europaea* × 1/2

[20] *Suaeda linearis* × 1⅕

[20a] *Suaeda linearis* × 1/4

[21] *Salsola kali* × 4/5

[22] *Amaranthus pumilus* × 1

[23] *Acnida cannabina* × 1/10

[24] *Froelichia floridana* × 1/4

[25] *Alternanthera philoxeroides* × 1/4

[26] *Iresine rhizomatosa* × 1/3

[27] *Philoxerus vermicularis* × 4/5

[28] *Phytolacca rigida* × 1/10

[29] *Mollugo verticillata* × 4/5

[31] *Portulaca oleracea* × 1⅛

[30] *Sesuvium portulacastrum* × 2⅓

[32] *Portulaca pilosa* × 4/5

[33] *Stellaria media* × 1

[34] *Stellaria graminea* × 3/5

[35] *Cerastium glomeratum* × 1

[36] *Cerastium fontanum* spp. *triviale* × 1⅕

[37] *Sagina procumbens* × 1⅕

[38] *Arenaria lateriflora* × 1/2

[39] *Honkenya peploides* × 1⅕

[40] *Spergularia rubra* × 1⅙

[41] *Odontonychia corymbosa* × 4/5

[43] *Silene pratensis* × 1/3

[42] *Scleranthus annuus* × 1

[44] *Saponaria officinalis* × 1/3

[45] *Brasenia schreberi* × 4/5

[46] *Nymphaea odorata* × 1/4

[47] *Nuphar luteum* × 1/2

[48] *Ranunculus septentrionalis* × 3/4

[49] *Ranunculus acris* × 1/4

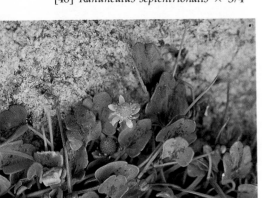

[50] *Ranunculus cymbalaria* × 1⅕

[51] *Ranunculus sceleratus* × 1/3

[52] *Argemone albiflora* × 3/5

[53] *Corydalis micrantha* × 4/5

[54] *Lepidium virginicum* × 3/4

[55] *Coronopus didymus* × 1

[56] *Cakile edentula* ssp. *harperi* × 3/5

[57] *Raphanus raphanistrum* × 1/5

[59] *Sarracenia purpurea* × 1/4

[58] *Barbarea vulgaris* × 1/5

[60] *Drosera rotundifolia* × 7/10

[61] *Fragaria virginiana* × 1/2

[62] *Potentilla simplex* × 1/3

[63] *Potentilla argentea* × 3/5

[64] *Potentilla recta* × 2/5

[65] *Mimosa strigillosa* × 1/2

[66] *Cassia aspera* × 2/5

[67] *Cassia fasciculata* × 2/5

[68] *Cassia obtusifolia* × 1/3

[69] *Baptisia tinctoria* × 1/20

[71] *Crotalaria rotundifolia* × 1

[70] *Crotalaria spectabilis* × 1/10

[72] *Lupinus perennis* × 1/4

[73] *Medicago polymorpha* × 3/4

[74] *Melilotus indica* × 2/5

[75] *Trifolium arvense* × 1¹⁄₁₀

[76] *Trifolium aureum* × 3/5

[77] *Trifolium hybridum* × 2/5

[78] *Trifolium repens* × 1/10

[79] *Trifolium pratense* × 3/4

[80] *Sesbania macrocarpa* × 1/3

[81] *Sesbania vesicaria* × 1/12

[82] *Aeschynomene indica* × 2/5

[83] *Stylosanthes biflora* × 1⅕

[84] *Desmodium paniculatum* × 1/3

[85] *Lespedeza capitata* × 1/2

[86] *Lespedeza repens* × 9/10

[87] *Vicia angustifolia* × 1/3

[88] *Vicia acutifolia* × 1⅗

[89] *Vicia dasycarpa* × 1/5

[90] *Lathyrus japonicus* × 1/3

[91] *Clitoria mariana* × 1/3

[92] *Centrosema virginianum* × 1/6

[93] *Erythrina herbacea* × 1/4

[94] *Galactia volubilis* × 1/3

[95] *Galactia elliottii* × 2/5

[96] *Rhynchosia minima* × 3/5

[97] *Strophostyles helvola* × 4/5

[99] *Geranium carolinianum* × 3/5

[98] *Vigna luteola* × 1/4

[100] *Oxalis stricta* × 1/3

[101] *Linum medium* × 4/5

[102] *Polygala grandiflora* × 3/4

[103] *Polygala lutea* × 1/4

[104] *Polygala incarnata* × 2

[105] *Croton glandulosus* × 1/6

[106] *Acalypha gracilens* × 1/3

[107] *Cnidoscolus stimulosus* × 1/3

[108] *Chamaesyce polygonifolia* × 9/10

[109] *Chamaesyce maculata* × 1³⁄₁₀

[110] *Chamaesyce nutans* × 2/5

[111] *Chamaesyce hirta* × 3/5

[112] *Euphorbia cyparissias* × 3/4

[113] *Impatiens capensis* × 1/2

[114] *Sida rhombifolia* × 1/2

[115] *Hibiscus moscheutos* × 1/3

[116] *Kosteletzkya virginica* × 1/2

[117] *Melochia corchorifolia* × 3/5

[118] *Hypericum perforatum* × 4/5

[119] *Hypericum gentianoides* × 1/2

[120] *Hypericum mutilum* × 1

[121] *Triadenum virginicum* × 3/5

[122] *Helianthemum carolinianum* × 3/5

[123] *Lechea racemulosa* × 1

[124] *Viola lanceolata* × 3/4

[125] *Passiflora incarnata* × 3/4

[126] *Rotala ramosior* × 1/2

[127] *Lythrum salicaria* × 1/5

[128] *Lythrum lineare* × 3/4

[129] *Rhexia alifanus* × 4/5

[130] *Rhexia nashii* × 2/5

[131] *Rhexia virginica* × 2/5

[132] *Ludwigia alternifolia* × 1/2

[133] *Ludwigia peruviana* × 2/5

[134] *Ludwigia palustris* × 4/5

[135] *Ludwigia linearis* × 2/5

[136] *Ludwigia virgata* × 1/2

[137] *Oenothera drummondii* × 1⅕

[138] *Oenothera humifusa* × 1/2

[139] *Oenothera laciniata* × 3/4

[140] *Oenothera biennis* × 1/7

[141] *Oenothera fruticosa* × 1/4

[142] *Oenothera speciosa* × 1/3

[143] *Gaura angustifolia* × 1⅕

[144] *Proserpinaca pectinata* × 1⅕

[145] *Aralia nudicaulis* × 1/5

[146] *Hydrocotyle bonariensis* × 1/2

[147] *Hydrocotyle umbellata* × 1¹/₁₀

[148] *Centella asiatica* × 1/3

[149] *Chaerophyllum tainturieri* × 1/2

[150] *Cicuta mexicana* × 1/5

[151] *Ptilimnium capillaceum* × 1/2

[152] *Lilaeopsis chinensis* × 1

[153] *Ligusticum scothicum* × 2/5

[154] *Heracleum maximum* × 3/8

[155] *Daucus carota* × 1/16

[156] *Monotropa uniflora* × 2/5

[157] *Samolus parviflorus* × 3/5

[158] *Lysimachia quadrifolia* × 1/4

[159] *Lysimachia terrestris* × 1/3

[175] *Matelea carolinensis* × 3/4

[176] *Cuscuta gronovii* × 3/4

[177] *Dichondra carolinensis* × 1

[178] *Calystegia sepium* × 1/4

[179] *Ipomoea brasiliensis* × 1/4

[180] *Ipomoea stolonifera* × 1/4

[181] *Ipomoea sagittata* × 1/3

[182] *Ipomoea trichocarpa* × 3/10

[183] *Ipomoea pandurata* × 1/7

[184] *Phlox drummondii* × 3/5

[185] *Heliotropium curassavicum* × 2/5

[186] *Verbena bonariensis* × 4/5

[188] *Phyla nodiflora* × 1/2

[187] *Verbena halei* × 1⅕

[189] *Teucrium canadense* × 1/4

[190] *Trichostema dichotomum* × 1⅖

[191] *Scutellaria epilobiifolia* × 3/5

[192] *Scutellaria integrifolia* × 2/5

[193] *Prunella vulgaris* × 4/5

[194] *Galeopsis tetrahit* × 1/3

[195] *Lamium amplexicaule* × 1⅛

[196] *Stachys floridana* × 3/5

[197] *Salvia lyrata* × 1/5

[198] *Salvia coccinea* × 1/2

[199] *Monarda punctata* × 3/5

[200] *Lycopus americanus* × 1/2

[201] *Hyptis alata* × 1/3

[202] *Physalis viscosa* L. ssp. *maritima* × 3/5

[203] *Solanum caroliniense* × 7/10

[204] *Solanum sisymbriifolium* × 3/10

[205] *Solanum dulcamara* × 2/5

[206] *Solanum pseudogracile* × 2/5

[207] *Verbascum thapsus* × 1/8

[208] *Linaria canadensis* × 1¹⁄₁₀

[210] *Gratiola virginiana* × 3/4

[209] *Linaria vulgaris* × 1½₀

[211] *Gratiola pilosa* × 1/2

[212] *Bacopa caroliniana* × 1

[213] *Bacopa monnieri* × 1⅙

[214] *Micranthemum umbrosum* × 1/2

[215] *Veronica arvensis* × 1/2

[216] *Veronica serpyllifolia* × 3/4

[217] *Agalinis fasciculata* × 3/4

[218] *Aureolaria flava* × 1/4

[219] *Buchnera americana* × 3/4

[220] *Utricularia inflata* × 1/4

[221] *Utricularia subulata* × 3/4

[222] *Utricularia purpurea* × 1/12

[223] *Plantago major* × 1/5

[224] *Plantago virginica* × 3/10

[225] *Plantago aristata* × 2/5

[226] *Plantago lanceolata* × 1/12

[227] *Hedyotis caerulea* × 3/5

[228] *Hedyotis procumbens* × 1⅕

[229] *Hedyotis uniflora* × 1³⁄₁₀

[230] *Richardia brasiliensis* × 9/10

[231] *Diodia teres* × 2/5

[232] *Diodia virginiana* × 1⅙

[233] *Galium aparine* × 2/5

[234] *Galium tinctorium* × 1/3

[235] *Galium hispidulum* × 3/5

[236] *Melothria pendula* × 3/4

[237] *Triodanis perfoliata* × 3/4

[238] *Lobelia cardinalis* × 1/4

[239] *Lobelia elongata* × 1/3

[240] *Elephantopus tomentosus* × 1/6

[241] *Eupatorium perfoliatum* × 1/4

[242] *Eupatorium capillifolium* × 1/8

[243] *Eupatorium aromaticum* × 1/4

[244] *Eupatorium rotundifolium* × 1/5

[245] *Eupatorium album* × 1/4

[246] *Eupatorium hyssopifolium* × 1/3

[247] *Eupatorium dubium* × 1/5

[248] *Mikania scandens* × 2/5

[249] *Liatris graminifolia* × 3/10

[250] *Carphephorus odoratissimus* × 1/7

[251] *Chrysopsis gossypina* × 1/3

[252] *Chrysopsis graminifolia* × 1/5

[253] *Chrysopsis falcata* × 2/5

[254] *Heterotheca subaxillaris* × 1/4

[255] *Solidago sempervirens* × 1/10

[256] *Solidago odora* × 2/5

[257] *Solidago canadensis* × 2/5

[258] *Solidago rugosa* × 1/2

[259] *Solidago uliginosa* × 2/5

[260] *Euthamia tenuifolia* × 3/8

[261] *Aphanostephus skirrhobasis* var. *thallassius* × 1/10

[262] *Aster tenufolius* × 2/5   [263] *Aster subulatus* × 3/4

[264] *Aster novi-belgii* × 1/4

[265] *Aster divaricatus* × 1/10

[266] *Aster lateriflorus* × 1/4

[267] *Aster pilosus* × 1/3

[268] *Aster linariifolius* × 3/5

[269] *Aster paternus* × 1/3

[270] *Erigeron quercifolius* × 2/5

[271] *Erigeron vernus* × 1/5

[272] *Erigeron myrionactis* × 1¾

[273] *Erigeron strigosus* × 1/10

[274] *Conyza canadensis* × 1³⁄₁₀

[275] *Conyza bonariensis* × 1/5

[276] *Pluchea odorata* × 1/3

[277] *Pluchea foetida* × 1/3

[278] *Pterocaulon pycnostachyum* × 3/10

[279] *Gnaphalium chilense* × 1/12

[280] *Gnaphalium obtusifolium* × 3/10

[281] *Gnaphalium purpureum*
var. *purpureum* × 1/3

[282] *Ambrosia artemisiifolia* × 1/4

[283] *Xanthium strumarium* × 1/4   [284] *Eclipta prostrata* × 1⅛

[285] *Rudbeckia hirta* × 1/3

286] *Helianthus angustifolius* × 1/5

[287] *Helianthus debilis* × 1/2

[288] *Melanthera nivea* × 1/3

[289] *Verbesina occidentalis* × 1/5

[290] *Verbesina virginica* × 1/3

[291] *Coreopsis tinctoria* × 1/4

[292] *Bidens laevis* × 2/5

[293] *Bidens bipinnata* × 3/10

[294] *Bidens pilosa* × 4/5

[295] *Bidens frondosa* × 2/5

[296] *Helenium amarum* × 1/5

[297] *Helenium autumnale* × 3/4

[298] *Gaillardia pulchella* × 1/8

[299] *Matricaria maritima* × 1/6

[300] *Achillea millefolium* × 1/8

[301] *Leucanthemum vulgare* × 1/4

[302] *Artemisia stelleriana* × 2/5

[303] *Erechtites hieracifolia* × 1/4

[304] *Senecio tomentosus* × 1/7

[305] *Cirsium horridulum* × 3/10

[306] *Cirsium vulgare* × 1/2

[307] *Cichorium intybus* × 3/4

[308] *Krigia virginica* × 3/5

[309] *Taraxacum officinale* × 1/4

[310] *Leontodon autumnalis* × 1/2

[311] *Sonchus asper* × 1/7

[312] *Lactuca canadensis* × 1/5

[313] *Pyrrhopappus carolinianus* × 1/3

[314] *Hieracium venosum* × 1/4

[316] *Hieracium caespitosum* × 4/5

[315] *Hieracium aurantiacum* × 1⅓

[317] *Hieracium gronovii* × 1/7

[318] *Triglochin striata* × 3/4

[319] *Sagittaria latifolia* × 1/4

[320] *Sagittaria lancifolia* × 3/8

[322] *Xyris iridifolia* × 1⅓

[321] *Sagittaria graminea* × 1/7

[323] *Eriocaulon decangulare* × 1⅙

[323a] *Eriocaulon decangulare* × 1/12

[324] *Tillandsia usneoides* × 1⅛

[325] *Tillandsia setacea* × 1/7

[327] *Tradescantia virginiana* × 3/5

[326] *Commelina erecta* × 1/2

[329] *Pontederia cordata* × 1/4

[328] *Eichhornia crassipes* × 1/3

[330] *Juncus roemerianus* × 3/5

[331] *Juncus effusus* × 1/2

[332] *Juncus gerardi* × 2/5

[333] *Juncus tenuis* × 3/5

[334] *Juncus megacephalus* × 1/10

[335] *Juncus validus* × 1/7

[336] *Allium canadense* × 3/4

[337] *Nothoscordum bivalve* × 1⅖

[338] *Lilium philadelphicum* × 2/5

[339] *Smilacina stellata* × 3/10

[340] *Maianthemum canadense* × 3/4

[341] *Hypoxis hirsuta* × 1/4

[342] *Iris versicolor* × 1/2

[343] *Sisyrinchium albidum* × 1/5

[344] *Sisyrinchium atlanticum* × 1/5

[345] *Sisyrinchium exile* × 1⅔

[346] *Canna flaccida* × 1/4   [347] *Cypripedium acaule* × 1/7

[348] *Spiranthes vernalis* × 1/2

[349] *Spiranthes cernua* × 3/8

[350] *Corallorhiza wisteriana* × 1/4

[351] *Epidendrum conopseum* × 1/4

[352] *Typha latifolia* × 1⅛

[353] *Tripsacum dactyloides* × 3/10

[354] *Erianthus giganteus* × 1/25

[355] *Andropogon glomeratus* × 1/25

[356] *Andropogon capillipes* × 1/9

[357] *Andropogon longiberbis* × 1/12

[358] *Andropogon ternarius* × 1/3

[359] *Sorghastrum secundum* × 1/100

[360] *Sorghum halepense* × 1/7

[361] *Schizachyrium littorale* × 1/11

[362] *Paspalum floridanum* × 1/3

[363] *Paspalum notatum* × 1/13

[364] *Paspalum distichum* × 1⅛

[365] *Paspalum urvillei* × 1/4

[366] *Panicum virgatum* × 1/30

[367] *Panicum amarum* × 1/5

[368] *Panicum rhizomatum* × 1/7

[369] *Panicum dichotomiflorum* × 1/3

[370] *Sacciolepis striata* × 3/5

[371] *Echinochloa walteri* × 1/4

[372] *Setaria glauca* × 1/3

[373] *Setaria geniculata* × 4/5

[374] *Setaria viridis* × 1/4

[375] *Setaria magna* × 1/9

[376] *Cenchrus tribuloides* × 3/4

[377] *Cenchrus echinatus* × 1

[378] *Cenchrus incertus* × 1⅙

[379] *Stenotaphrum secundatum* × 1/2

[380] *Zizania aquatica* × 1/9

[381] *Muhlenbergia filipes* × 1/50

[382] *Sporobolus virginicus* × 1/4

[383] *Sporobolus indicus* × 3/8

[384] *Polypogon monspeliensis* × 3/5

[385] *Agrostis stolonifera* × 1/4

[386] *Ammophila breviligulata* × 1/20

[387] *Holcus lanatus* × 1/2

[388] *Spartina cynosuroides* × 1/5

[389] *Spartina alterniflora* × 1/5

[390] *Spartina patens* × 1/4

[391] *Eustachys petraea* × 5/8

[392] *Eleusine indica* × 1/4

[393] *Triplasis purpurea* × 1/13

[394] *Phragmites australis* × 1/4

[395] *Eragrostis spectabilis* × 1/5

[396] *Uniola paniculata* × 1/40

[397] *Chasmanthium sessiliflorum* × 1/6

[398] *Distichlis spicata* × 3/5

[399] *Briza minor* × 3/5

[400] *Poa annua* × 1⅛

[401] *Vulpia octoflora* × 7/8

[402] *Elymus virginicus* × 1/4

[403] *Cyperus esculentus* × 1/4

[404] *Cyperus strigosus* × 1/10

[405] *Cyperus haspan* × 2/5

[406] *Cyperus retrorsus* × 2/5

[407] *Cyperus filicinus* × 1/2

[408] *Cyperus brevifolius* × 3/4

[409] *Fuirena pumila* × 1/2

[410] *Fuirena scirpoidea* × 1⅛

[411] *Scirpus americanus* × 1⅕

[412] *Scirpus robustus* × 1⅗

[413] *Scirpus cyperinus* × 1/8

[414] *Eleocharis tuberculosa* × 1/8

[415] *Eleocharis flavescens* × 1⅙

[416] *Fimbristylis castanea* × 3/5

[417] *Dichromena latifolia* × 3/5

[418] *Cladium jamaicense* × 1/16

[419] *Rhynchospora corniculata* × 4/5

[420] *Scleria triglomerata* × 1/2

[421] *Carex crinita* × 1/3

[422] *Pinus elliottii* × 1/8

[423] *Pinus echinata* × 1/4

[424] *Pinus glabra* × 1/4

[425] *Pinus rigida* × 1/4

[426] *Chamaecyparis thyoides* × 2/5

[427] *Juniperus virginiana* × 2/5

[428] *Sabal palmetto* × 1/80

[429] *Serenoa repens* × 1/20

[430] *Yucca gloriosa* × 1/30

[431] *Yucca aloifolia* × 1/11

[433] *Smilax auriculata* × 3/8

[432] *Yucca flaccida* × 1/16

[434] *Smilax bona-nox* × 1/3

[435] *Smilax glauca* × 3/8

[436] *Smilax laurifolia* × 1/4

[437] *Smilax pumila* × 1/4

[438] *Populus alba* × 2/5

[439] *Populus grandidentata* × 1/5

[440] *Salix caroliniana* × 1/3

[441] *Salix bebbiana* × 1/3

[442] *Myrica cerifera* × 2/5

[443] *Myrica pensylvanica* × 1/2

[444] *Myrica gale* × 1

[445] *Comptonia peregrina* × 4/5

[446] *Carya glabra* × 1/4

[447] *Carya tomentosa* × 9/10

[448] *Carya illinoensis* × 1/7

[449] *Ostrya virginiana* × 2/5

[450] *Corylus americana* × 2/5

[451] *Betula papyrifera* × 3/5

[451a] *Betula papyrifera* × 1/160

[452] *Betula nigra* × 3/8

[453] *Alnus serrulata* × 1/4

[454] *Fagus grandifolia* × 1/4

[455] *Castanea pumila* var. *ashei* × 2/5

[456] *Quercus nigra* × 1/4

[457] *Quercus marilandica* × 1/4

[458] *Quercus coccinea* × 1/4

[459] *Quercus velutina* × 1

[460] *Quercus falcata* × 1/4

[461] *Quercus laevis* × 1/3

[462] *Quercus ilicifolia* × 3/5

[463] *Quercus myrtifolia* × 3/4

[464] *Quercus laurifolia* × 1/5

[465] *Quercus incana* × 2/5

[466] *Quercus alba* × 1/4

[467] *Quercus stellata* × 1/3

[468] *Quercus chapmanii* × 1/2

[470] *Quercus virginiana* × 1/2

[469] *Quercus geminata* × 1/2

[471] *Ulmus americana* × 1⅕

[472] *Celtis laevigata* × 1/4

[473] *Morus rubra* × 2/5

[474] *Phoradendron serotinum* × 1/4

[475] *Sarcocornia perennis* × 1/2

[476] *Batis maritima* × 1/2

[477] *Magnolia grandiflora* × 1/4

[478] *Magnolia virginiana* × 4/5

[479] *Asimina parviflora* × 2/5

[480] *Cinnamomum camphora* × 3/8

[481] *Persea borbonia* var. *borbonia* × 2/5

[482] *Sassafras albidum* × 1/4

[483] *Ribes hirtellum* × 1

[484] *Liquidambar styraciflua* × 1/4

[485] *Hamamelis virginiana* × 1/3

[486] *Spiraea latifolia* × 5/8

[487] *Aronia arbutifolia* × 1⅛

[487a] *Aronia arbutifolia* × 3/8

[488] *Malus angustifolia* × 3/4

[489] *Amelanchier arborea* × 1/2

[490] *Amelanchier laevis* × 5/8

[491] *Crataegus brainerdii* × 1/3

[492] *Rubus trivialis* × 1/5

[493] *Rubus cuneifolius* × 1/2

[494] *Rubus strigosus* × 5/8

[495] *Rosa rugosa* × 1/2

[496] *Rosa virginiana* × 3/8

[497] *Rosa palustris* × 2/5

[498] *Prunus serotina* × 3/8

[499] *Prunus virginiana* × 3/8

[500] *Prunus pensylvanica* × 3/5

[501] *Prunus umbellata* × 2/5

[502] *Prunus caroliniana* × 1/4

[503] *Licania michauxii* × 1/3

[505] *Cercis canadensis* × 2/5

[504] *Acacia smallii* × 3/4

[506] *Amorpha fruticosa* × 1/4

[507] *Robinia pseudoacacia* × 1/5

[508] *Daubentonia punicea* × 3/10

[509a] *Xanthoxylum clava-herculis* × 1/3

[509] *Xanthoxylum clava-herculis* × 1/6

[510] *Croton punctatus* × 2/5

[511] *Sapium sebiferum* × 2/5

[512] *Corema conradii* × 1³⁄₁₀

[513] *Ceratiola ericoides* × 4/5

[514] *Rhus copallina* × 1/6

[515] *Rhus typhina* × 1/4

[516] *Toxicodendron radicans* × 1/4

[517] *Toxicodendron vernix* × 1/8

[518] *Cliftonia monophylla* × 3/5

[520] *Ilex opaca* × 1/5

[519] *Cyrilla racemiflora* × 1/4

[521] *Ilex vomitoria* × 4/5

[522] *Ilex glabra* × 1/3

[523] *Ilex verticillata* × 2/5

[524] *Ilex cassine* × 1/5

[525] *Acer rubrum* × 2/5

[526] *Aesculus pavia* × 1/4

[527] *Berchemia scandens* × 1/5

[528] *Sageretia minutiflora* × 2/5

[529] *Vitis aestivalis* × 1/3

[530] *Vitis labrusca* × 1/4

[532] *Parthenocissus quinquefolia* × 3/10

[531] *Vitis rotundifolia* × 1/3

[533] *Ampelopsis arborea* × 1/3

[534] *Hypericum hypericoides* × 3/4

[535] *Hypericum tetrapetalum* × 3/5

[536] *Hypericum cistifolium* × 1/3

[537] *Tamarix* spp. × 3/4

[538] *Helianthemum corymbosum* × 1/10

[539] *Hudsonia tomentosa* × 3/4

[540] *Opuntia humifusa* × 1/3

[541] *Opuntia pusilla* × 1/5

[542] *Decodon verticillatus* × 1/12

[543] *Aralia spinosa* × 1/6

[544] *Nyssa biflora* × 1/4

[545] *Cornus florida* × 1/3

[546] *Cornus amomum* × 3/10

[547] *Cornus alternifolia* × 1/2

[548] *Clethra alnifolia* × 3/5

[549] *Kalmia angustifolia* × 2/5

[550] *Lyonia ferruginea* × 1/3

[551] *Lyonia ligustrina* × 3/5

[552] *Lyonia lucida* × 2/5

[553] *Cassandra calyculata* × 1/2

[555] *Gaultheria procumbens* × 7/10

[554] *Epigaea repens* × 3/5

[556] *Arctostaphylos uva-ursi* × 1⅖

[557] *Gaylussacia baccata* × 7/10

[558] *Gaylussacia dumosa* × 9/10

[559] *Vaccinium angustifolium* × 4/5

[560] *Vaccinium corymbosum* × 3/5

[561] *Vaccinium arboreum* × 3/10

[562] *Vaccinium myrsinites* × 2/3

[563] *Vaccinium macrocarpon* × 7/10

[564] *Vaccinium stamineum* × 1/2

[565] *Bumelia tenax* × 1/2

[566] *Diospyros virginiana* × 1/3

[567] *Symplocos tinctoria* × 3/10

[568] *Osmanthus americanus* × 1/6

[569] *Forestiera segregata* × 3/5

[570] Gelsemium sempervirens × 1/2

[571] Lantana camara × 1/2

[572] Callicarpa americana × 3/10

[573] Avicennia germinans × 1/2

[574] *Conradina canescens* × 4/5

[575] *Lycium carolinianum* × 1¹⁄₁₀

[576] *Bignonia capreolata* × 1/4

[577] *Campsis radicans* × 1/4

[578] *Cephalanthus occidentalis* × 3/10

[579] *Sambucus canadensis* × 1/10

[580] *Viburnum recognitum* × 2/5

[581] *Viburnum cassinoides* × 1/3

[582] *Lonicera japonica* × 1/2

[583] *Lonicera sempervirens* × 3/10

[584] *Baccharis halimifolia* × 1/4

[585] *Baccharis angustifolia* × 3/10

[586] *Iva frutescens* × 1/3

[587] *Iva imbricata* × 3/4

[588] *Borrichia frutescens* × 1/4

Descriptions

Herbaceous Dicots

shrubs or trees or in open, flats, marshes; eTX-ME. July–Oct.

SAURURACEAE
Lizard's-tail Family

Lizard's-tail
Saururus cernuus L.

[1]
Perennial, often forming extensive colonies, spreading by rhizomes as well as seeds. Perianth absent. Stamens 3–7. Carpels 3–4, nearly separate. The spike of flowers droops at the tip but the spike becomes erect as the seeds mature. Nothing else similar in aquatic habitats. Common. Swamps, marshes, margins of streams and ponds, low woodlands; eTX–RI. Apr.–July.

URTICACEAE
Nettle Family

False-nettle
Boehmeria cylindrica (L.) Sw.

[2]
Perennial without stinging hairs, to 1.3 m tall. Leaf blades to 15 cm long and 7 cm wide. Flowers small, lacking petals, unisexual, on axillary spikes; fruits are small achenes that are enclosed by the enlarged calyx. Common. Moist or wet soil—under

POLYGONACEAE
Buckwheat Family

Sheep Sorrel
Rumex acetosella L.

[3]
A weedy perennial to 50 cm tall; from slender, running rootstocks. Leaves 2–5 cm long, generally hastate, the 2 basal lobes usually divergent. Flowers unisexual on separate plants. Sepals at fruiting stage just equaling achene and tightly enclosing it. Common. Open places—stable dunes; interdunes; fields, pastures; NC–Greenl. Apr.–Oct.

Wild Sorrel
Rumex hastatulus Baldw. ex Ell.

[4]
Annual or short-lived perennial without rhizomes, to 1.2 m tall. Stems single or in large clumps, often forming extensive and colorful colonies in open areas. Leaves usually with 2–4 widely divergent lobes. Flowers unisexual and sexes on separate plants. Sepals at fruiting stage expanded into broad wings much wider than the achene. Common. In open—dunes, interdunes, roadsides, fields, thin woods; eTX–NC. Mar.–Aug.

Swamp Dock; Water Dock
Rumex verticillatus L.

[5]
Perennial to 1.5 m tall, from a deep tap root. Leaves entire, essentially flat. Fruiting pedicels 2-5 times as long as fruiting sepals. Mature sepals wider than achene, midveins of all 3 mature sepals with swollen bases that project below the sepals. Common. Marsh and pond edges, swamps, sloughs; eTX–LA; SC–NBr. Apr.–July.

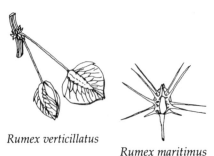

Rumex verticillatus

Rumex maritimus

Two other species are easily recognized by long bristles on the mature sepals. Both are annuals, often bushy branched, and often have crumpled leaf margins. On one, *R. maritimus* L. (Golden Dock), bristles are clearly longer than width of sepals. Occasional. Saline or brackish marshes or shores; NJ–NS. July–Aug. On the other, *R. persicarioides* L. (Seashore Dock), bristles are about as long as width of sepals. Rare. Similar places; MA–sNS. July–Sept.

Curly Dock, Sour Dock
Rumex crispus L.

[6]
Stout, erect perennial to 1.5 m tall. Leaves crumpled and puckered, especially near margins. Pedicels usually less than twice as long as fruit. Mature sepals entire, all 3 midveins notably enlarged at base. Common. Sandy shores, marshes, swales, public areas; TX–Nfld. Mar.–Aug.

Water Dock
Rumex orbiculatus Gray

[7]
Stout perennial to 2.5 m tall. Leaves essentially flat. Mature sepals entire or nearly so, with swollen portion of midribs distinctly above the pedicel. Occasional. Brackish flats, between stable dunes, swamps, shores; NJ–Nfld. June–Aug.

Rumex orbiculatus

R. *pallidus* Bigel. (Pale Dock) is a slender branching perennial easily recognized by glaucous, narrowly lanceolate leaves (basal ones sometimes wider) and the swollen midveins of the mature sepals being almost as long as the sepals. Rare. Saline or brackish marshes; NY–Nfld. June–July.

Knotweed
Polygonum aviculare L.

[8]
Much branched annual, prostrate to erect, not glaucous. Stipules forming sheaths around stem above nodes. Leaves small, narrow to the base. Flowers in small clusters in leaf axils. Sepals oblong to ovate-oblong. Fruit an achene 2–3 mm long. Common northward, rare southward. Along shores, yards, along trails; SC–Nfld. June–Nov.

Polygonum glaucum

P. glaucum Nutt. (Seaside Knotweed) is of the same general appearance but is very glaucous, sepals obovate and narrow at base, achene 3–4 mm long. Occasional. Sandy beaches, brackish swales, minidunes, edge of salt marshes; FL–MA. July–Nov.

Water Smartweed
Polygonum punctatum Ell.

[9]
Perennial to 1 m tall, perhaps annual in the north. Stems long, decumbent, and rooting. Racemes slender and usually interrupted in lower part. Tip of sheathing stipules with bristles that sometimes fall from old stipules. Calyx white to greenish white, with glandular dots. Fruits are black shiny achenes. Common. Swamps, freshwater or brackish marshes, swales, sloughs, ponds; TX–NS. June–frost.

Polygonum punctatum

Polygonum punctatum

P. densiflorum Meisn. also is a perennial and has a glandular-dotted calyx but the sheathing stipules lack bristles at summit and the racemes are more compact. Rare. Usually in water—swamps, freshwater marshes, swales. NC–VA. June–frost.

Pinkweed
Polygonum pensylvanicum (L.) Small

[10]
Abundantly branching annual to 2 m tall, sheathing stipules thin with no bristles at tip, tending to break apart. Flowers pink or rarely white; in plump, blunt, mostly straight racemes. Fruits are dark brown to black shiny achenes. Common. Fresh or brackish shores, swales, ditches, disturbed areas; eTX–MS; nGA–wNS. June–frost.

*Polygonum
persicaria*

*Polygonum
pensylvanicum*

P. persicaria L., also an annual, has similar flowers, leaves, and achenes. It may be separated by bristles on summit of the sheathing stipules. Occasional. Similar habitats; FL–NS. June–frost.

Water-pepper
Polygonum hydropiperoides Michx.

[11]
Somewhat slender perennial to 50 cm tall. Stems often decumbent and rooting. Sheathing stipules with bristles on summit and stiff ascending appressed hairs on sides. Racemes slender. Calyx white to pink, rarely greenish or purplish, without glandular dots. Common. Swamps, freshwater marshes, ponds, swales, sloughs, margins of brackish places, ditches; TX–eNS. May–frost.

*Polygonum
hydropiperoides*

*Polygonum
setaceum*

P. setaceum Baldw. ex Ell. is similar but usually coarser, to 1.5 m tall, and sheathing stipules have loose spreading hairs on sides. Occasional. Similar places; TX–NY. May–frost.

Tearthumb
Polygonum sagittatum L.

[12]
Retrorsely barbed, abundantly branched, slender-stemmed annual. Leaves lanceolate to elliptic, sagittate at base, 3–8 cm long, sharply retrorsely barbed on midvein below. Flowers in long-stalked heads or interrupted spikes. Common. Wet places in open—fresh or brackish sloughs and ponds; GA–Nfld. July–frost.

   P. arifolium L. is similar but may be recognized by hastate leaves 10–12 cm long or occasionally longer and not retrorsely barbed below. Occasional. Fresh or brackish tidal marshes, wet meadows; eFL–NBr. July–frost.

Climbing False-buckwheat
Polygonum convolvulus L.

[13]
Twining annual. Stems sharply angled. Leaves acute to acuminate, cordate to cordate-sagittate. Flowers in axillary racemes. Calyx without a wing on keel. Fruit dull achenes. Rare. Roadsides, edges of parking lots, fields, disturbed places; SC–Greenl. May–Nov.

Polygonum convolvulus

Polygonum scandens

   P. scandens L. is quite similar but a perennial, with shiny achenes and wing-keeled calyx. Common. Usually moist places—thickets, fresh and brackish marshes; VA–NS. July–Nov.

Jointweed
Polygonella articulata (L) Meisn.

[14]
Erect annual to 60 cm tall. Branches appearing to rise from internodes. Leaves linear, blades 1–3 cm long, jointed at junction of summit of sheathing stipule. Flowers abundant, on upper ⅔ of plant. Pedicels jointed above the base. Sepals 5, white or greenish to dark pink, separate above, united at base into a stalklike structure (stipe) longer than calyx lobes. Common. Open dry sandy areas; VA–sME. Aug.–Nov.

Polygonella articulata

   P. gracilis (Nutt.) Meisn. is more slender and the flowers not as crowded. The stipe is shorter than the calyx lobes. Occasional, active and stable dunes, interdune areas, thin scrub; eLA–sGA. Aug.–Nov.

CHENOPODIACEAE
Goosefoot Family

Lamb's-quarters
Chenopodium album L.

[15]
Annual to 3 m tall. Leaves lanceolate to broadly ovate or deltoid; not glandular, but young ones white-mealy. Flowers very small, in dense clusters, perfect. Petals none. Sepals 4–5, fleshy, without wings or spines, persistent, tightly enclosing fruit. Seeds

1 per fruit, horizontal, black, shiny, 1–1.5 mm long; the thin fruit cover persistent when rolled between hands. Common. Stable dunes, interdunes, edge of fresh and brackish marshes, fields, waste places; TX–Nfld. June–frost.

C. rubrum L. has similar flower clusters. It differs in having main leaves glabrous, rhombic-ovate to oblong, and with a conspicuous tooth on each side near base, sometimes other teeth above. Young leaves not white-mealy. Seeds mostly vertical. Rare. Salt and brackish marshes, brackish sloughs; NJ–Nfld. Aug.–frost. Other species of the genus are unlikely to be encountered.

Wormseed; Mexican-tea
Chenopodium ambrosioides L.

[16]
Usually an annual but plants in south may live several years. Leaves flat, with many sessile glands, strongly aromatic. Flowers very small, perfect, in dense glomerules on short to long spikes, the spikes usually many in large terminal groups. Petals absent. Sepals 5, fleshy, persistent, tightly enclosing fruit. Source of chenopodium oil, which is used to treat domestic stock for intestinal worms. Common. Edge of fresh or brackish marshes, interdunes, swales, beaches; TX–ME. July–frost.

Sea-beach Atriplex
Atriplex arenaria Nutt.

[17]
Annual with angled stems to 50 cm long. Leaves silvery gray, flat, alternate, scurfy with branlike scales, ovate and rounded to narrowly lance-ovate and tapering at base. Flowers

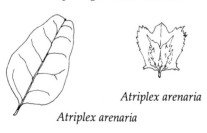

Atriplex arenaria

Atriplex arenaria

unisexual. Staminate flowers in terminal spikelike structures; pistillate in axillary clusters, lacking a calyx, completely enclosed by a pair of bracts. Fruiting bracts with angular teeth about 1 mm long across the broad tip. Common. Seashores, active dunes, minidunes, overwash areas, edge of brackish or saline marshes; TX–NH. July–frost.

Sea-beach Atriplex
Atriplex patula L.

[18]
Annual with angled stems to 1 m long. Leaves flat, green to purplish green; lower opposite, blades truncate to hastate at base, generally triangular to rhombic, apex obtuse; upper leaves lanceolate to narrowly ovate or oblanceolate, the base tapering. Triangular teeth on the tapering tip of fruiting bracts under 0.5 mm

Atriplex patula

long or lacking. Vegetatively often confused with some *Chenopodium* species but clearly separated by the fruiting bracts. Common. Saline or brackish places—overwash areas, sloughs, marshes, edges of ponds; GA–Greenl. July–frost.

Glasswort
Salicornia europaea L.

[19]
In *Salicornia* foliage leaves are absent but represented by small opposite appressed scales. Flowers small, inconspicuous, deeply sunk in scale axils.

This species is an erect annual to 50 cm tall with scales below reproductive spikes mucronate. Fruiting spikes 1.5–3 mm thick. Plants often occur in conspicuous colonies, and usually turn brilliant red in autumn. Common. Upper portions of brackish or salt marshes, low overwash areas; eFL–Nfld. July–Oct.

In *S. virginica* L., also an annual, scales are blunt or rounded, fruiting spikes 4–6 mm thick. Plants are usually greenish in autumn. Common. Similar places; TX–ME. June–Oct. Syn.: *S. bigelovii* Torr. The perennial *S. virginica* of current manuals is presented in the section on woody plants under the genus *Sarcocornia*.

Sea-blite
Suaeda linearis (Ell.) Moq.

[20, 20a]
Suaeda spp. have glabrous, entire, linear, alternate, fleshy leaves with no spiny tips; flowers sessile, in groups of 3 in leaf axils; corolla absent; calyx glabrous and closely curving over fruit.

This species is an annual, or sometimes perennial in south, to 80 cm tall, sometimes reclining but not mat-forming. Leaves are green, not glaucous. Sepals unequal, the larger 3 hoodlike on the back. Seeds 1–1.5 mm wide. Common. Salt or brackish marshes, overwashes, sea beaches; TX–sME. Aug.–frost.

There are 2 similar species, occurring in similar places. One, *S. maritima* (L.) Dum. is erect to forming mats, leaves usually glaucous, sepals equal, seeds about 2 mm wide. Common. FL; VA–NS. Aug.–frost. In *S. richii* Fern. plants are prostrate and form mats, sepals equal; seeds 1–1.5 mm wide. Rare. MA–Nfld. Sept.–frost.

Bassia hirsuta (L.) Aschers is much like *Suaeda* but the stem, leaves, and calyx are hairy. Occasional. VA–MA. July–Sept.

Russian-thistle
Salsola kali L.

[21]
Bushy-branched annual, green to bright red. Leaves alternate, awl-shaped, very sharp-pointed. Flowers perfect, in leaf axils. Common. Along seashores, small active dunes, margins of saline or brackish marshes, dry sandy soils; TX–Nfld. June–frost.

AMARANTHACEAE
Amaranth Family

Beach Amaranth
Amaranthus pumilus Raf.

[22]
Annual with fleshy, prostrate to decumbent stems to 40 cm long, often branching abundantly and forming mats. Leaves fleshy, crowded toward tip of stems, blades scarcely longer than wide. Flowers in short axillary clusters, unisexual, both sexes on same plant. Sepals dry, membranous. Fruit fleshy, indehiscent, 4–5 mm long. Seed 1, elliptic, 2–2.5 mm long. Occasional. Beaches and low active dunes, often covered by tides; SC–RI and Nantucket I. June–frost.

Some other members of the genus occur in seaside habitats, but rarely. They are weedy erect annuals to 2 m tall, difficult to identify to species.

Water-hemp
Acnida cannabina L.

[23]
Annual with stout stems, to 3 m tall, or flowering when only 10–20 cm tall when beginning growth late in the season, inflorescences often turning red in autumn. Leaves alternate and petioled, blades to 20 cm long, lanceolate to ovate-lanceolate or nearly oblong, the apex obtuse or rounded. Inflorescence branches spikelike, stiff. Flowers unisexual, male and female on separate plants. Bracts below male flowers about 1 mm long, in the female flowers less than 1 mm. Fruit 2.5–4 mm long, indehiscent. Common in north to rare in south. Salt to fresh tidal marshes and shores; FL–sME. July–frost. Syn.: *Amaranthus cannabina* (L.) Sauer.

A. australis Gray is similar but sometimes taller. Bracts of all flowers 1.5–1.8 mm long. Fruit 2 mm long or less. Occasional. Similar places; TX–wFL. July–frost. Syn.: *Amaranthus australis* (Gray) Sauer.

Cottonweed
Froelichia floridana (Nutt.) Moq.

[24]
Annual to 1.8 m tall. Stem solitary or more commonly with a few well-developed erect branches from the base or above, internodes progressively longer upward. Stem and leaves hairy, grayish to brownish. Leaves opposite, entire, oblanceolate to spatulate or oblong. Flowers in terminal spikes 10–12 mm in diameter. Petals absent. Sepals 5, densely woolly, united at base. Stamens 5, the filaments united into a 5-lobed tube exceeding the anthers. Fruit woolly, 5 mm long. Occasional. Sandy soils— stable dunes, interdunes, pinelands, fields; TX–sNS. June–Oct.

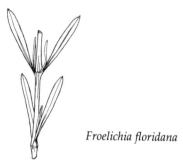

Froelichia floridana

Separated into 2 varieties by some: var. *floridana* with leaf blades widest near the middle, var. *campestris* (Small) Fern. with blades widest above middle.

Alligator-weed
Alternanthera philoxeroides (Mart.) Griseb.

[25]
An especially noxious weed in places, conducive to mosquito breeding. Perennial with decumbent stems to 1 m long, often forming tangled masses in water. Leaves opposite, lanceolate to elliptic, not succulent, to 9 cm long, tipped with a tiny spine. Flowers in headlike spikes on leafless stalks to 6 cm long. Stamens 5, united at base into a tube. Freshwater habitats— ponds, sloughs, depressions, ditches; TX–sVA. Apr.–frost.

Iresine
Iresine rhizomatosa Standl.

[26]
Stoloniferous perennial to 1.5 m tall. Leaves opposite, to 15 cm long. Distinctive when in flower. Flowers very small, numerous, in terminal panicles, male and female flowers separate, on different plants. This species is related to *I. herbstii* Hook. f. (Bloodleaf), a common house and flower garden plant. Occasional. Swales, sloughs, edge of salt or brackish marshes, damp woods; eTX–MD. Aug.–frost.

Silverhead
Philoxerus vermicularis (L.) R. Br.

[27]
Perennial. Stems prostrate, branching, and rooting abundantly, often forming mats; or occasionally ascending to 50 cm tall. Leaves opposite, thick, fleshy, to 6 cm long. Flowers in dense spikes, perfect, filaments united into a short cup, stigmas 2, other flower parts of 5 each. Occasional. Saline or brackish places— coastal dunes, damp shores, flats, lagoon shores, margins of marshes; TX–wFL. Flowers entire year in southernmost parts of coast.

PHYTOLACCACEAE
Pokeweed Family

Pokeweed
Phytolacca rigida Small

[28]
Glabrous perennial to 3 m tall. Leaves pliable, with the feel of thin kid leather. Flowering and fruiting racemes permanently erect or nearly so and appearing to be attached laterally on the stems but actually attached terminally. Fruits a dark purple 5- to 12-carpelled berry. Roots and older parts of tops poisonous when eaten. Common. Stable dunes, interdunes, disturbed places; eTX–VA. May–Oct.
   P. americana L. is similar but has

nearly all racemes divergent to declined; the first one to develop and occasionally a few others are erect, especially in the flowering stage. Common. Freshwater situations—waste areas, disturbed places, fencerows, recent clearings; eTX–sME. May–Oct.

AIZOACEAE
Carpetweed Family

Carpetweed: Indian-chickweed
Mollugo verticillata L.

[29]
Annual with small prostrate to ascending stems. Leaves in apparent whorls of 3–6, mostly unequal in size. Flowers 2–5 from each node. Petals none. Sepals 5, about 2 mm long, green with white margins. Fruit a 3-carpelled many-seeded capsule. Seeds dark reddish brown with several parallel ridges. Common. Weedy—stable dunes, swales, edge of brackish and freshwater marshes and ditches, gardens, fields, pastures, yards; TX–NS. Mar.–frost, or all year in south.

Sea-purslane
Sesuvium portulacastrum L.

[30]
Succulent sprawling much-branched perennial rooting at nodes. Leaves 1–6 cm long, opposite, one of the pair often larger than the other, tips pointed. Flowers solitary, appearing axillary, on pedicels over 3 mm long. Petals none. Stamens many. Fruit a capsule, splitting horizontally at base, freeing the single seed. Seed about 1.5 mm long. Common. Beaches, drift areas, overwash areas, minidunes, brackish swales, upper parts of salt marshes; TX–sNC. Frost-free periods.

　S. maritimum (Walt.) B.S.P. is similar but is an annual, stems mostly erect or if prostrate not rooting, leaves usually relatively wider, flowers sessile or nearly so, sepals 3–5

mm long, seed about 0.8 mm long. Common in south, rare in north. Similar places; TX–NY. Frost-free periods.

PORTULACACEAE
Purslane Family

Common Purslane
Portulaca oleracea L.

[31]
A decumbent to prostrate much-branched conspicuously fleshy annual. Leaves alternate and opposite, spatulate to obovate. Fruit with a lid opening just below the middle at an even circular line. Seeds dark red or black, 0.6–1 mm broad. Horticultural varieties have been developed with flowers 12–30 mm across. Used as food in India for over 2000 years and in Europe for hundreds. Frequently offered for sale in Mexican markets. Occasional. Upper beaches, drift areas, edge of brackish marshes, soils of dried pools; cultivated areas, waste places; TX–NS. Apr.–frost.

Hairy Portulaca
Portulaca pilosa L.

[32]
Annual to 20 cm tall, much branched. Hairs in leaf axils and at base of flowers. Petals to 6 mm long. Stamens 15 or more. Fruit with a lid opening near middle in an even circular line, the base persisting as seen in picture. Seeds many, reddish-black. Occasional. Dry sandy soils—between stable dunes, along roads, exposed shores of ponds, yards, waste places; LA–sNC. May–Nov.

　Selected plants with showy flowers to 2 cm across are used in flower gardens. These have been designated as a variety, *Portulaca pilosa* var. *hortualis* Bailey.

CARYOPHYLLACEAE
Pink Family

Common Chickweed
Stellaria media (L.) Vill.

[33]
Members of this family have 4–5 persistent sepals, an equal number of petals, and 1-celled capsular fruits with many seeds and central placentation (or indehiscent 1-seeded fruits in a few genera).

In this genus leaves lack stipules, sepals are separate, petals appear to be 10 but are actually only 5 and deeply notched, styles usually 3, and capsule not cylindrical.

This species is a highly variable opposite-leaved annual with weak stems. Plants usually compact at first but later loosely branched and often forming dense masses. Stems hairy in lines. Leaves petioled, blades ovate. Sepals and some other parts are from glandular-hairy to glabrous. Many species of birds are known to use chickweed for food. Common. Between stable dunes, edge of marshes, fields, gardens, yards, waste places, meadows, roadsides; eTX–Greenl. Flowering all year in south, May–frost in north.

Common Stitchwort
Stellaria graminea L.

[34]
Perennial with weak 4-angled glabrous stems. Leaves linear-lanceolate to lanceolate, widest below the middle. Flowers in well-developed terminal and axillary peduncled clusters with small thin nongreen bracts. Seeds 0.8–1.2 mm long, finely roughened. Occasional. Moist to wet places—between old dunes, grassy places, roadsides; nNC–Nfld. May–Oct.

The obviously related *S. longifolia* Muhl. ex Willd. has linear long-tapering leaves that are widest at the middle, flowers nearly all in long-peduncled axillary clusters, seeds smooth. Rare. Similar places; MA–Nfld. May–Aug.

Mouse-ear Chickweed
Cerastium glomeratum Thuill.

[35]
Cerastium species are annual or short-lived perennials with opposite leaves, no stipules, separate sepals, petals 2-lobed or shallowly cleft; capsule many-seeded, frequently curved, opening at end by 10 teeth.

This species is an erect viscid-hairy annual with ovate to obovate leaves, the lower spatulate. Flower clusters compact or becoming open in age, but terminal ones always dense. Inflorescence bracts green throughout and without thin margins. Pedicels shorter than capsules. Capsule less than twice as long as sepals. Common. Stable dunes and interdunes, along roads, waste places; eTX–sNfld. Feb.–July. Syn.: *C. viscosum* L.

Large Chickweed
Cerastium fontanum ssp. *triviale* (Link) Jalas

[36]
Short-lived matted viscid-hairy perennial with many flowering stems to 50 cm tall. Leaves conspicuously hairy. Inflorescence open, not compact, bracts with clear margins outside the greenish interior portion. Petals about as long as sepals. Capsule 7–11 mm long. Common. Disturbed areas, fields, around buildings, roadsides; nNC–NS. Mar.–June; autumn; sporadically in summer. Syn.: *C. vulgatum* L.; *C. holosteoides* var. *vulgare* (Hartm.) Hylander.

Another matted perennial that may be encountered is *Cerastium arvense* L. It has ascending to erect flowering branches to 40 cm tall. Bracts have clear margins. Petals are 2–3 times as long as sepals. Rare. Similar places, often mixed with grasses. DE–Greenl. Apr.–June.

Birdseye
Sagina procumbens L.

[37]
Sagina species have opposite narrowly linear leaves with no stipules. Sepals separate. Petals separate, entire, and

not lobed; or absent. Styles as many as sepals. Fruit a many-seeded capsule with 4–5 teeth.

This species is a matted perennial to 15 cm tall. Sepals 4 (rarely 5). Petals 0–5. Capsule 2–3.5 mm long. Seeds 0.2–0.3 mm long, very finely pebbled. Common except in south. Damp open areas—swales, edge of brackish marshes, roadside ditches; NJ–Greenl. May–frost.

Although an annual, *S. decumbens* (Ell.) T. & G. is quite similar except not matted, sepals 5 (rarely 4) and seeds with fine slender ridges. Stems often many and in tufts. Common. Shrub zone in dunes, edge of brackish marshes, depression in paths and roads; TX–sMA. Mar.–June.

Sandwort
Arenaria lateriflora L.

[38]
Perennial with slender rhizomes, often forming colonies. Leaves opposite, ovate to elliptic-ovate, tip usually obtuse, with no stipules. The 5 sepals shorter than the 5 entire petals. Styles 3. Capsule many-seeded, 4–7 mm long, slenderly conic to ovoid, splitting into 3 deeply 2-parted sections. Common. Edge of brackish marshes, shrub thickets, woods, swales between stable dunes; NY–Nfld; circumboreal. May–Aug. Syn.: *Moehringia lateriflora* (L.) Fenzl.

Sea-beach Sandwort
Honkenya peploides (L.) Ehrh.

[39]
Distinctly fleshy perennial. Stems much-branched deep in sand as well as above, forming dense masses to 2 m wide, stems at perimeter usually prostrate. Sepals separate. Capsule separating into 3–5 sections. Seeds few, 3–4.5 mm long. Occasional, rare in south. Sandy beaches, between small dunes; MD–Nfld, then circumboreal. May–Aug. Syn.: *Arenaria peploides*.

Sand Spurrey
Spergularia rubra (L.) J. & C. Presl

[40]
Spergularia species are annuals or short-lived perennials with opposite leaves and stipules. Capsule splitting to base in 3 parts.

This species is usually prostrate and often matted, has clusters of leaves in axils of the primary leaves that are strongly mucronate. Stipules much longer than wide. Petals pink. Stamens 6–10. Seeds brown, 0.4–0.6 mm long. Occasional. Weedy, sandy or gravelly soils, waste places, pans; MD–Nfld. May–Oct.

S. marina (L.) Griseb. is usually much-branched, erect or nearly prostrate. Leaves mucronate. Stipules about as long as wide. Petals white or pink. Stamens 2–5. Seeds 0.6–0.8 mm long. Common. Saline or brackish soils, flats, sea beaches; TX–NS. Apr.–Oct. *S. canadensis* (Pers.) Don is similar but leaves blunt and seeds 0.8–1.4 mm long. Occasional. Similar places; NY–Nfld. July–Sept.

Whitlow-wort
Odontonychia corymbosa (Small) Small

[41]
Perennial with finely hairy stems, opposite narrow entire leaves, long conspicuous stipules. Flowers in compact clusters, bracts below flowers falling at maturity. Base of flower minutely hairy. Petals none, sepals long and narrow, with no short abrupt point. Fruit a utricle about 1 mm long. Occasional. Dunes and beaches; MS–wFL. Apr.–Aug. Syn.: *Paronychia erecta* var. *corymbosa* (Small) Chaudhri.

O. interior Small is similar but not as coarse and base of flower is bristly hairy. Rare. Dunes, coastal sands, shrub; wFL. June–Oct. Syn.: *Paronychia rugelii* Schuttlew. ex Chapm. var. *interior* (Small) Chaudhri. In the similar *O. erecta* (Chapm.) Small stems are glabrous. Occasional. Coastal sands and dunes, shrub; MS–wFL. June–Sept. Syn.: *Paronychia erecta* var. *erecta*.

Knawel
Scleranthus annuus L.

[42]
Much-branched annual to 35 cm in diameter, often forming conspicuous mats. Leaves 3–20 mm long, under 1 mm wide, lacking stipules. Flowers green, sessile in leaf axils. Calyx forming a cup 1–2 mm long, lobes to 2 mm long, the whole longer than the fruit. Corolla absent. Fruit a 1-seeded utricle. Common. Dry interdunes, paths, roadsides, waste places; MS–Nfld. Frost-free periods.

White Campion
Silene pratensis (Rafn) Godron & Gren.

[43]
Annual to about 1 m tall, sometimes persisting for a second year. Leaves opposite, lanceolate to broadly elliptic, to 15 cm long. Flowers unisexual, male and female on separate plants. Sepals united, the lobes much shorter than the tube, tubular when in flower, ovoid to nearly globose in fruit. Corolla white. Styles 4–5. Fruit a several-seeded capsule. Occasional. Roadsides, around buildings, fields, waste places; NC–NS. May–Sept. Syn.: *S. alba* (Mill.) Krause; *Lychnis alba* Mill.

Soapwort; Bouncing-bet
Saponaria officinalis L.

[44]
Glabrous perennial, spreading from seed and rhizomes. Stems to 1.5 m long, decumbent to erect. Leaves opposite. Sepals united into a cylindrical, or nearly so, tube that is obscurely 5-nerved, the lobes about 2 mm long. Petals separate, white to light pink. Styles 2. The scientific name comes from sapo, soap. The mucilaginous juice forms a lather with water and has been used as a soaplike material since ancient Greek times. Plant may be poisonous if eaten. Occasional, rare in south. Roadsides, fencerows, fields, waste places; TX–NS. May–Oct.

CABOMBACEAE
Water-shield Family

Water-shield
Brasenia schreberi J. F. Gmel.

[45]
Perennial with slim rhizomes creeping in mud. Leaf blades peltate, elliptic, not notched, to 11 cm long. Stems, petioles, and undersurface of leaves with a prominent mucilaginous cover. Sometimes mistaken for a water-lily because of its leaves but the flower is unlike those of native water-lilies. Occasional. In or at edge of fresh water—ponds, sloughs; FL–NS. June–frost.

NYMPHAEACEAE
Water-lily Family

Water-lily; Water-nymph
Nymphaea odorata Ait.

[46]
Perennial from a large rhizome. Leaf blades floating or lying on mud, purple beneath, nearly orbicular, notched at the base. Sepals 4, nearly separate. Stamens fastened all over sides of the ovary. Seeds maturing under water. Flowers fragrant. Petals white or rarely pinkish. Common. In or at edge of fresh water—ponds, sloughs, marshes, swamps; eTX–Nfld. Apr.–Oct.
   N. mexicana Zucc. is similar but the petals are yellow. Rare. Similar places; TX–SC.

Yellow Pond-lily
Nuphar luteum (L.) Sibth. & Sm.

[47]
Perennial from a large rhizome. Leaves submersed, floating, or emersed. Leaf blades ovate to orbicular, deeply cut at base. Stamens many, fastened under the ovary. Common. In or at edge of fresh water, tidal areas, ponds, sloughs, swamps; eTX–Nfld. Apr.–Oct.

RANUNCULACEAE
Buttercup Family

Swamp Buttercup
Ranunculus septentrionalis Poir.

[48]
Members of this genus are herbaceous annuals or perennials with basal or alternate leaves or both, and 3–5 green or greenish sepals. Petals 5, yellow or rarely white, each with a nectar-bearing pit or scale on the upper side at the base. Stamens many, yellow, attached directly beneath the several to many greenish pistils. Fruits achenes. There are many species and most plants are difficult to name. Fortunately for our purposes, few species are likely to be encountered along seacoasts. Several species are known to be poisonous and all should be considered potentially so.

This species is perennial; develops trailing, often rooting, stems; has 3-parted basal leaves with the terminal segment or all segments stalked. Achenes not turgid, minutely pitted, margin thin. Rare along coast but common inland. Wet places in woods or in open—swales, swamps, pebbly shores; NJ–RI. Apr.–July.

Tall Buttercup
Ranunculus acris L.

[49]
Conspicuous perennial with 1 to few erect stems to 1.5 m tall. Basal leaves reniform, deeply palmately cleft, terminal segment sessile. Sepals spreading, 4–7 mm long. Petals 6–16 mm long. Fruiting head 5–7.5 mm thick. Achenes 2–3.5 mm long, faces obliquely obovate, and very finely dotted. Common. Weedy, moist to dry places—protected beaches, freshwater marshes, fields, pastures, thin woods; NY–Greenl. May–Aug.

R. bulbosus L. (Bulbous Buttercup) is also conspicuous and is superficially similar but the 1 to several stems only to 65 cm tall. The base is distinctively bulbous, to 3.5 cm wide. Terminal segment of the leaves is

Ranunculus bulbosus

Ranunculus acris

stalked, sepals reflexed, fruiting head 7.5–10 mm thick. Occasional. Weedy, often in poorly drained areas—roadsides, grasslands, around buildings, thin woods; nNC–Nfld. Apr.–June.

Seaside Buttercup
Ranunculus cymbalaria Pursh

[50]
Perennial herb to 25 cm tall, spreading by filiform stolons. Leaves cordate to ovate or reniform, crenate. Petals about 3.5 mm long. Achenes 1.5–2.3 mm long. Common. Saline or brackish places—marshes, swales; NJ–Greenl. May–Oct.

Cursed Buttercup
Ranunculus sceleratus L.

[51]
Erect annual to 60 cm tall. Basal leaves mostly deeply palmately 3-parted, the 3 parts cleft again. Petals 1.5–3 mm long. Achene turgid, 0.8–1.4 mm long, beak very minute, faces smooth to faintly roughened horizon-

tally. Occasional. Moist to wet places—edge of brackish pools, edge of marshes, tidal mud, roadside ditches, swamps, swales, wet meadows; TX–Nfld. Mar.–Aug.

R. *abortivus* L., also an annual, is of similar stature, occasionally taller (to 85 cm), with similar flowers and achenes, but the basal leaves are reniform with usually cordate bases, or a few lobed or divided, the margins crenate. Occasional. Usually moist places under other plants or in open—edge of freshwater marshes, swamps, thickets, roadsides, fields; NY–Nfld. Mar.–Aug.

PAPAVERACEAE
Poppy Family

Prickly Poppy
Argemone albiflora Hornem.

[52]
Short prickly annual with large showy flowers, thus easily recognized. Leaves coarsely cleft or parted and with spiny teeth. Sepals spiny, 2–3. Petals 3–6 cm wide. Stamens many. Fruit a capsule opening at the top by 3–5 sections. Occasional. Waste places, roadsides, dunes, interdunes, overwash grasslands, fence-rows; TX–NC. Mar.–June.

A. *mexicana* L. is similar but the petals are yellow. Vegetative parts and seeds of this species are known to be poisonous. Rare. Similar places; TX–sNC.

FUMARIACEAE
Fumitory Family

Slender Corydalis
Corydalis micrantha (Engelm.) Gray

[53]
Green or glaucous annual or biennial. Stems erect to decumbent, branching mostly from the base. Sepals 2, small. Corolla yellow, sometimes with light

reddish longitudinal lines, 9–15 mm long including the spur on one of the petals. Fruit a 1-celled capsule with 2 parietal placentas. Occasional. Dry or moist usually sandy soils—drift zone at edge of marshes, roadsides, cultivated areas, waste places; TX–NC. Feb.–May.

BRASSICACEAE
Mustard Family

Peppergrass; Pepperwort
Lepidium virginicum L.

[54]
Glabrous or minutely hairy annual or biennial to 90 cm tall; the plant illustrated is unusually small. Basal leaves incised to pinnate. Petals 4, equal, white. Stamens 2. Fruits 3–3.5 mm long, faintly notched at tip, 2-celled, flattened perpendicular to partition, orbicular, smooth. Seeds 1–2 per fruit. Common. Dry open soil, salt spray areas, between stable dunes, fields, gardens, yards; TX–Nfld. Apr.–Nov.

L. *campestre* (L.) R. Br. is similar but densely short hairy, fruits 5–6 mm long, stamens 6. Occasional. Similar places; NC–Nfld. Mar.–June. In L. *latifolium* L. the fruits are not notched at apex and about 2 mm long. Stamens 6. Rare. Beaches, tidal shores; NY–MA. May–June.

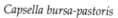

Capsella bursa-pastoris

Lepidium campestre

Lepidium virginicum

Capsella bursa-pastoris (L.) Medic. (Shepherd's-purse) is a plant of similar growth habit but with basal leaves more prominent and with distinctive triangular-obcordate fruits 5–10 mm long. Petals 2–4 mm long. Occasional. Weedy—roadsides, yards, disturbed places; TX–Greenl. Feb.–Oct.

Wart-cress; Carpet-cress
Coronopus didymus (L.) Sm.

[55]
Prostrate to ascending annual, often forming mats, distinctive because of its fetid odor and fruits that are notched on both ends to the extent that the two 1-seeded halves are nearly globose. Fruits 2.5–3 mm wide and about 1.5 mm long, the surface deeply wrinkled. Leaves deeply pinnately lobed. Common. Yards, fields, disturbed habitats, waste places, roadsides, cultivated areas; eTX–Nfld. Feb.–Oct.

Sea-rocket
Cakile edentula (Bigel.) Hook. ssp. *harperi* (Small) Rodman

[56]
Cakile species are succulent annuals with flowers in terminal racemes. Sepals 4, green. Petals 4, white to pink or purple. Fruits are unusual, distinctly divided into 2 segments of different shape, the basal one persistent and 1-seeded or seedless. Terminal segment gradually tapers to a short beak, usually 1-seeded, and eventually breaks off. When mature, the terminal segment is dry and corky and can float great distances; the lower usually falls later and does not go far.

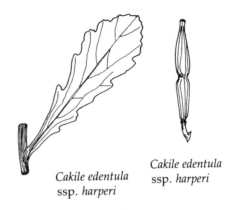

Cakile edentula
ssp. *harperi*

Cakile edentula
ssp. *harperi*

This subspecies has a linear inflorescence, mature pedicels not as broad as the axis of the inflorescence, fruit 4–7 mm wide and the upper segment 8-ribbed when dry. Common. Sea beaches, low active dunes, overwash areas, edge of salt marshes; eFL–NC. Mar.–Oct. Subspecies *edentula* is similar but the dried upper segment is 4-angled. Common. Similar places. NC–Nfld. Mar.–Sept. *C. constricta* Rodman has fruits only 3–4 mm wide. Common. Sandy shores; eTX–FL. Mar.–Aug. The inflorescence is geniculate and the fruiting pedicels as broad as the axis in *C. geniculata* (Robins.) Millsp. Common. Sandy drift areas; TX–LA. Mar.–Aug.

Jointed Charlock; Wild Radish
Raphanus raphanistrum L.

[57]
Hairy annual with a large taproot. Petals 10–15 mm long, yellow, sometimes white when old. Mature pedicels ascending. Fruit elongate, indehiscent, the body 2–4 cm long and the beak 1–3 cm long; at maturity constricted between the 4–10 seeds. Possibly poisonous when eaten. Common in north, rare and local in the south. Drift areas of beaches and marshes, overwash areas, fields, waste places; SC–Greenl. Mar.–Nov.

Cakile constricta

Cakile edentula
ssp. *edentula*

Raphanus raphanistrum

Winter-cress
Barbarea vulgaris R. Br.

[58]
Annual or biennial with glabrous stems, at least in the upper part. Basal leaves petioled, with 1–4 pairs of small elliptic to ovate lateral lobes and a large rotund terminal lobe. Upper stem leaves sessile or clasping. Petals 4, yellow. Fruits linear, terete, dehiscent, 15–30 mm long, beak 1.8–3 mm long, seeds in 1 row in each of the 2 cavities. Occasional. Moist places—pastures, fields, roadsides, waste places; VA–Nfld. Apr.–June.

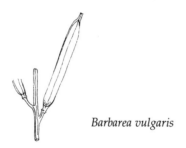

Barbarea vulgaris

SARRACENIACEAE
Pitcher-plant Family

Pitcher-plant
Sarracenia purpurea L.

[59]
Perennial with decumbent hollow leaves, containing liquid in varying quantities. The inner surface of the nearly flattened leaf tip bears many stiff hairs that point away from the opening in the leaf. Insects trapped by these hairs are often digested, thus providing some of the plant's food and mineral needs. Some small organisms such as mosquito larvae find a suitable home in the leaves. The flowers are strange in that they are all nodding and the end of the pistil is expanded into a large persistent 5-lobed umbrellalike structure. This species is the provincial flower of Newfoundland. Occasional in north, rare in south. Peaty soil of bogs, thin wet pinelands and savannas; NJ–Nfld. May–Aug.

DROSERACEAE
Sundew Family

Sundew
Drosera rotundifolia L.

[60]
Drosera species have stalked glands on the leaves. These glands exude a clear sticky substance that traps insects.

In this species leaves are all basal and the flowering stems are glabrous. Leaf blades nearly round but most slightly broader than long. Petals white to pink, 5–7 mm long. Common except in south. Margins of ponds and sloughs, peaty soil of bogs, moist savannas, swales; VA–Greenl. June–Sept.

D. capillaris Poir. is similar but leaf blades are obovate to spatulate, the petioles with a few long hairs and about as long as or a little longer than the blades. Occasional. Similar places; eTX–sVA. Apr.–Aug. In *D. brevifolia* Pursh the petioles are glabrous. Common. Similar places; eTX–sSC. Mar.–May. Syn.: *D. leucantha* Shinners. In *D. intermedia* Hayne leaves are not all basal, and the petiole is 2–6 cm long, much longer than the blades. Occasional. Similar places; TX–FL; NC–Nfld. May–Sept.

ROSACEAE
Rose Family

Strawberry
Fragaria virginiana Duchn.

[61]
Perennial with long stolons. All leaves basal, compound, leaflets toothed along most of margins. Flowers in peduncled clusters, sepals alternating with entire bracts, petals white. Fruit an enlarged red receptacle with many small achenes (from separate carpels) sunken in shallow pits on the surface. Many people prefer the flavor of wild strawberries over cultivated strains. Quite variable with several varieties named. Common northward, rare in south. Barren fields, thin woods, stable dunes; NC–Nfld. Mar.–July.

In *F. vesca* L. the achenes are not in pits. Rare. Rich soils, woodlands, roadsides; NY–Nfld. Apr.–July. Leaves and fruits of *Duchesnea indica* (Andr.) Focke (Indian Strawberry) are similar to those of *Fragaria*. The fruits are essentially tasteless and bracts that alternate with the sepals are conspicuous and toothed. Petals yellow. Occasional. Usually in moist and shady places—waste places, lawns, around buildings; eTX–CT. Feb.–frost.

Old-field Cinquefoil
Potentilla simplex Michx.

[62]
Members of this genus are herbs and shrubs. Leaves on various parts of a plant are mostly similar. Entire bracts alternate with the sepals. Petals yellow, rarely white or purple. Carpels many, styles short. Ovaries superior, the receptacle little enlarged in fruit. Fruits are small achenes clustered tightly on the dry receptacle. Perennial with a basal rosette of palmately compound leaves. Leaflets 5, sharply serrate on the upper two-thirds of the blade. Stems nearly erect at first, later arching and often rooting at tips, up to 1 m long. Flowers solitary, the first one usually from the node above the second well-developed internode. Common. Open, dry to moist places—thin woods, pastures, roadsides, edge of bogs; SC–NS. Apr.–July.

P. canadensis L. is similar but the stems at first erect or ascending, later elongating and becoming prostrate. Leaflets serrate in upper half or less of the blades. First flower usually from the node above the first well-developed internode. Common. Dry open areas—thin woods, roadsides, disturbed areas, barren fields; VA–NS. Mar.–June.

Silvery Cinquefoil
Potentilla argentea L.

[63]
Perennial with freely branching stems that are nearly prostrate to ascending. Leaves palmately compound. Leaflets 5–7, the lower surface densely matted with silvery hairs. Flowers 7–15 mm wide, in terminal clusters, petals yellow and equalling to shorter than sepals. Achenes nearly smooth. Occasional. Dry open places, sandy areas behind and between stable dunes, roadsides, barren fields; MD–Nfld. May–Aug.

Rough-fruited Cinquefoil
Potentilla recta L.

[64]
Perennial with 1 to several erect stems. Leaves with 7–9, rarely 5, leaflets, the lower leaves abundant and on long hairy petioles. Flowers many, 1.5–2.5 cm wide, on erect stalks in a broad terminal cluster. Petals usually pale yellow, broadly notched at apex. Fruits seedlike, with low curved ridges. Common. Roadsides, rocky shores, waste places, usually dry places; NC–Nfld. Apr.–Sept.

P. norvegica L. is similar but the leaves have only 3 leaflets. Petals about as long as sepals. Common. Similar places; VA–Greenl. Apr.–Sept.

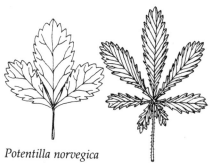

Potentilla norvegica

Potentilla recta

FABACEAE
Bean Family

Mimosa; Powderpuff
Mimosa strigillosa T. & G.

[65]
Perennial with spreading prostrate stems. Leaves alternate, pinnately twice-compound. Flowers in pink globose heads. Stamens 8–10. Fruits are bristly-hairy legumes 1–2 cm long. Occasional. Sandy soils in thin woods or open, grasslands, hammocks, roadsides; TX–LA; lower wFL–sGA. May–Sept.

Schrankia microphylla (Dryander) Macbr., also a prostrate or trailing perennial herb, has similar flowers and leaves but leaves and stems have numerous small curved prickles. The many leaflets "go to sleep" when disturbed, only midvein evident beneath. Common. Dry sandy soils in thin woods or open. MS–SC. May–Sept. *S. uncinata* Willd. is similar to this species but the lateral veins on the underside of the leaflets are prominently raised. Rare. Usually sandy soils in open; eFL–sGA. May–Sept.

Partridge-pea
Cassia aspera Muhl. ex Ell.

[66]
Annual with once-pinnate leaves. Stems, leaves, and fruits with conspicuous spreading hairs. Petiole with a small gland near lowest leaflets. Leaflets 7–25 pairs. Flowers 1–3 in leaf axils. Petals 4–7 mm long, slightly unequal. Stamens 7–9. Seeds of this and other partridge-peas are a very important source of food for wild birds. Common. Open often disturbed areas, pinelands, dunes, scrub areas; FL except nw, to sSC. May–Oct.

C. nictitans L. lacks the conspicuous spreading hairs, having small incurved hairs. Otherwise quite similar to *C. aspera*. Common. Roadsides, fields, pastures, sterile sandy soils, stable dune hollows, thin disturbed woods; eTX–MA.

Partridge-pea
Cassia fasciculata Michx.

[67]
Similar to *C. aspera* and *C. nictitans*, but larger. Stems nearly glabrous to conspicuously hairy, erect to prostrate. Leaflets 5–18 pairs. Flowers 2.5–4 cm across, corolla yellow to nearly white. Stamens 10. Common. Roadsides, thin pinelands, between stable dunes, disturbed areas; TX–sMA. June–Sept.

Coffee-weed; Sickle-pod
Cassia obtusifolia L.

[68]
Erect annual to 1.5 m tall. Leaflets 2–3 pairs, mostly 3–6 cm long and 2–4 cm wide, broadly obtuse, an elongate gland between or just below the 2 lower leaflets. Flower on pedicels over 10 mm long. Stamens 6–7, the upper 3 or 4 lacking normal anthers, the 3 lowest with conspicuous anthers. Fruit 15–45 mm long, very slender, usually strongly curved, 4-angled or nearly so. Important as alternate host of the tobacco etch virus disease. Recently proved to be toxic when eaten in considerable amounts. Common. Waste places, cultivated land, pastures, between stable dunes; eTX–NC. June–frost. Some books incorrectly refer to this species as *C. tora* L., which is an Old World species.

False-indigo; Rattleweed
Baptisia tinctoria (L.) R. Br.

[69]
Bushy-branched perennial to 1 m tall with alternate palmately compound nearly sessile leaves. Leaflets 3, obovate, cuneate. Flowers in numerous few-flowered racemes terminal on branches. Corolla yellow. Stamens 10, separate. Fruit an inflated pod 7–9 mm long and with a stalk between the pod and the calyx. Occasional. Dry, usually sandy soil—thin woods, clearings, heaths, stable dunes; VA–sME. Apr.–Sept.

Showy Crotalaria
Crotalaria spectabilis Roth

[70]
Members of this genus have 10 united stamens, the anthers alternately of 2 sizes, and stalkless inflated pod.

This species is a conspicuous annual to 1.5 m tall. Leaves simple, glabrous, mostly 6–20 cm long. Stipules and bracts at base of flower stalks ovate, 5 mm long or longer. Any part of the plant, especially the seeds, is poisonous when eaten. Growing plants are reported to reduce or eliminate nematodes from the soil. Occasional. Fields, roadsides, disturbed areas, stable dunes and interdunes; eTX–sSC. Mar.–Oct.

C. retusa L. is similar but the stipules and bracts are narrow and under 2 mm long. Rare. LA–wFL. July–Sept.

Rabbit-bells
Crotalaria rotundifolia (Walt.) Poir.

[71]
Perennial to 40 cm tall from a prominent taproot. Stems erect, several to many, with appressed hairs; or decumbent and with appressed or spreading hairs. Leaves simple, upper surface hairy, blades linear to orbicular. The inflated pod 12–28 mm long, 5–11 mm thick. Plants, especially seeds, poisonous when eaten. Common. Sandy areas, pinelands, thin woods, stable dunes; eLA–sNC. Mar.–Nov. Syn.: *C. angulata* Mill.
Other similar species are as follows:

C. sagittalis L. is erect to strongly ascending and has spreading hairs on the stems. Upper surface of leaf hairy. Common. Similar places; eLA–MS; VA–MA. May–Oct. *C. purshii* DC. with glabrous upper surface of leaf. Occasional. Sandy areas, pinelands, thin upland woods; MS–NC. Apr.–Aug.

Lupine
Lupinus perennis L.

[72]
Lupinus species have 10 stamens, the lower half of all filaments united into a tube. Anthers of 2 shapes, adjacent anthers different in shape. Fruits are flattened legumes. Many species poisonous when eaten.

This species has palmately compound leaves with 7–11 leaflets. Flowers in racemes to 30 cm long. Occasional. Open places in thin woods, swales between old dunes; VA–sME. Apr.–July.

L. villosus Willd. has a similar form but petals flesh-colored to deep pink and the center of the standard reddish purple. Leaves distinctive in being simple, the blade oblong to elliptic and evergreen. Fruits are shaggy with hairs 4–5 mm long. Rare. MS–FL; SC–sNC. Mar.–May.

Bur-clover; Medick
Medicago polymorpha L.

[73]
Annual with prostrate to ascending stems. Leaves compound; stipules deeply divided, mostly to below the middle. Leaflets 3, terminal one stalked, mostly longer than wide, finely serrate, without a center spot. Flowers 3–5 in a head. Petals yellow,

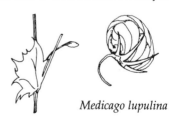

Medicago lupulina

Medicago arabica

Medicago lupulina   Medicago polymorpha

Medicago polymorpha

2.5–4 mm long. Fruits spirally coiled 2–5 times, with 2 rows of flexible prickles. Occasional. Weedy—yards, around buildings, roadsides, waste places; TX–SC. Mar.–Oct.

   M. arabica (L.) Huds. (Spotted Medick) is similar but leaflets have a purple blotch near center. Petals 4–5 mm long. Rare. Similar places; eTX–NC. Mar.–Aug. In *M. lupulina* L. (Black Medick) the stipules are acuminate, slightly toothed along margins. Petals yellow, 1.5–2 mm long. Pod kidney-shaped, l-seeded. Occasional. Similar places; TX–NS. Mar.–Sept.

Sour Clover
Melilotus indica (L.) All.

[74]
Spreading or ascending annual with alternate compound leaves. Leaflets 3, finely serrate, terminal leaflet stalked. Flowers 2–3 mm long, on pedicels under 1 mm long, in slender racemes. Corolla yellow. Fruits 1- to 2-seeded, about 2 mm long. Occasional. Weedy—roadsides, waste places, around buildings, back beaches, stable dunes areas; TX–sSC. Apr.–Oct.

   M. officinalis (L.) Pall. (Yellow Sweet Clover) also has yellow petals but the flowers are 5–7 mm long, pedicels 1.5–2 mm long. Used as hay, but

when molded can cause internal hemorrhaging and even death. Common. Fields, grasslands, roadsides, various soils; nGA–Nfld. Apr.–Oct. In *M. alba* Medic. (White Sweet Clover) the petals are white. Occasional. Fields, roadsides, waste places, especially calcareous areas; NC–Nfld. Apr.–Oct.

Rabbit-foot Clover
Trifolium arvense L.

[75]
Members of this genus have compound leaves with 3 finely serrate to dentate leaflets and 10 uniform-sized united stamens, one of these nearly free. Corolla irregular, withering and persisting, concealing the indehiscent ripened pod.

   This species is an annual; flower clusters unlike those of any other clover. They are peduncled, cylindrical, soft, and resemble a rabbit's foot. Corolla shorter than calyx. The stems branch freely and are topped by numerous heads in vigorous plants. If mature heads are eaten by livestock, they may cause mechanical intestinal irritation. Common locally. Dry fields, roadsides, waste places, swales, stable dunes, heaths; SC–NS. Apr.–Aug.

Hop Clover
Trifolium aureum Pollich

[76]
Annual or biennial to 50 cm tall. Terminal leaflet nearly sessile. Petals yellow. Occasional. Roadsides, waste places, fields, swales, stable dunes; VA–Nfld. June–Sept. *T. agrarium* L. as used by various manuals.

   Two similar species, both commonly called Low Hop Clover, are as

Trifolium dubium   Trifolium campestre

follows: *T. campestre* Schreb. (*T. procumbens* L.) has terminal leaflet distinctly stalked, heads 20- to 30-flowered, and flowers 3.5–4.5 mm long. Occasional. Similar places; eTX–NS. Mar.–Oct. *T. dubium* Sibth. has stalked terminal leaflets, but heads 3- to 15-flowered and flowers 2.5–3.5 mm long. Occasional. Similar places; TX–NS. Mar.–Sept.

Alsike Clover
Trifolium hybridum L.

[77]
Erect to ascending perennial 30–80 cm tall with glabrous stems. Flowers 30–50 in each head, the individual flowers 8–11 mm long and stalked. Corolla nearly white to washed with rose. Extensively planted for forage and escaping. Occasional. Fields, pastures, roadsides, around buildings, swales between stable dunes; FL–Nfld. Apr.–Oct.

White Clover
Trifolium repens L.

[78]
Glabrous, or nearly so, perennial with creeping stems that root at the nodes. Often forms large masses. Individual flowers pediceled, the long (10–25 cm) peduncled heads arising from the creeping stems. Calyx teeth are shorter than the tube. Common. Roadsides, yards, fields, thin woods, around buildings; TX–Nfld. Apr.–Oct.

Red Clover
Trifolium pratense L.

[79]
Perennial with several stems from a strongly developed taproot. Stipule tips triangular and abruptly awned. Flower heads globose, sessile. Flowers over 1 cm long and sessile. Extensively cultivated and escaping. Bumble bees are especially effective in pollinating Red Clover. Honey bees are also effective but will usually choose the more easily available nectar and pollen in the smaller flowers of other species such as sweet clovers. Common. Fields, roadsides, protected pebble and sand beaches, between dunes, other open places; NC–Nfld. Apr.–Sept.

In *T. incarnatum* L. (Crimson Clover) the petals are scarlet or deep red (rarely white) and the fully developed flower heads are peduncled and cylindrical. Overripe Crimson Clover can be dangerous to horses, as the fibrous calyx may become impacted in the digestive tract. Frequently cultivated for pasture and soil improvement, and escaped, especially in south. Common locally. Roadsides, fields, along railroads; TX–sME. Apr.–June.

Sesbania
Sesbania macrocarpa Muhl.

[80]
Annual to 4 m tall, mostly around 2 m, thinly branched. Leaves evenly pinnate, leaflets 20–70. Flowers in peduncled axillary clusters. Corollas shaped like those of sweet peas. Stamens 10, one filament separate from the 9 united into a tube. Fruit a linear glabrous legume, 10–23 cm long, 3–4 mm thick, and with cross partitions. Occasional. Thin low pinelands, ditches, edge of brackish and fresh marshes, swales, edge of sloughs, fields, alluvial soils; eTX–sNC. June–Oct. Syn.: *S. exaltata* (Raf.) Rydb.

Bladder-pod
Sesbania vesicaria (Jacq.) Ell.

[81]
Annual to 4 m tall. Leaves evenly pinnate. Flowers 8–15 mm long, in racemes shorter than the leaves. Petals yellow, tinged with red, the wings sometimes mostly dark pink. Fruit tapered at both ends, the body flattened; at maturity the firm outer layer separates from a thin soft layer enclosing the seeds. Seeds 2, rarely 1 or 3; poisonous, with consumption as low as 0.05 percent of body weight causing death. Cattle sometimes choose the fruits as food and hundreds from a single herd have been killed under particular circumstances.

Occasional. Low areas in open places, edge of marshes, between stable dunes, ditches; eTX–sNC. July–Sept. Syn.: *Glottidium vesicarium* (Jacq.) Harper.

Joint-vetch
Aeschynomene indica L.

[82]
Bushy annual to 2.5 m tall. Leaves even-pinnate, sometimes appearing odd-pinnate. Leaflets 19–70, not gland-dotted, with 1 major vein, this slightly off-center. Flowers 8–10 mm long. Corolla yellow. Stamens united by filaments into 2 groups of 5 each. Fruit with 5–12 joints, breaking into 1-seeded sections at maturity. These are used for food by birds. Occasional. Wet places in thin woods, edge of marshes, wet meadows, between stable dunes, ditches; TX–NC. July–Oct.

A. virginica (L.) B.S.P. is similar but the flowers are mostly over 10 mm long. Occasional. Tidal shores of rivers and marshes, thin woods in swamps; GA–sNJ. July–Oct. *A. viscidula* Michx. also has a yellow corolla and similar stamens but is sticky-haired and fruits with only 2–5 joints. Fruits can be confused with those of some *Desmodium* spp. but corollas not yellow in the latter. Occasional. Sandy pinelands and scrub; TX; MS–sGA. June–Sept.

Aeschynomene viscidula

Aeschynomene viscidula

Pencil-flower
Stylosanthes biflora (L.) B.S.P.

[83]
Perennial with 1 to several prostrate to erect stems. Stipules fastened to the petiole and also forming a sheath around the stem. Leaflets 3, lanceolate to oblanceolate, oval, or elliptic. Petals yellow. Fruit in two sections, the lower one usually aborting and becoming stipelike; commonly eaten by birds. Common. Thin dry woods, sandy soils; TX–VA. May–Sept.

Those plants that are prostrate to slightly ascending have been recognized by some as *S. riparia* Kearn. Recent studies, however, indicate that there is only one species.

Beggar-ticks
Desmodium paniculatum (L.) DC.

[84]
A world-wide genus with numerous species. Species here are trailing to erect, have 3 leaflets, and stipulelike structures (stipels) just below each leaflet. Fruits are distinctive, being loments (i.e., they break apart into segments). These readily cling to passing objects by means of hooked hairs. Abundantly eaten by wildlife, especially birds. We recorded 11 species in seaside habitats. Most are difficult to name. The 3 included here are fairly easy to recognize.

This species is a perennial to 1.2 m tall. Stipules 3–6 m long. Terminal leaflets 3–6 times as long as wide, stipels persistent. Petals 6–8 mm long. Fruit usually of 2–6 triangular or nearly so segments. Common. Thin woods, between stable dunes, sandy fields and meadows; eTX–NY. May–Oct.

If stems are trailing and flowers 4–5.5 mm long, plants are *D. lineatum* DC. Occasional. Similar places; eTX–VA. May–Aug. The only annual to be encountered is *D. tortuosum* (Sw.) DC. It is erect, to 2 m tall, loment with 2–7

Desmodium lineatum

Desmodium lineatum

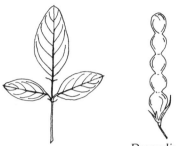

Desmodium tortuosum   *Desmodium tortuosum*

oval to suborbicular segments. Occasional. Similar places; AL–sNC. June–Aug.

Lespedeza
Lespedeza capitata Michx.

[85]
Lespedezas have pinnately 3-foliate leaves, entire leaflets. Fruits indehiscent, 1-seeded, and not adherent. All species provide food for animals, the foliage as forage, the fruits abundantly eaten, especially by birds. Although the genus is not difficult to recognize, most species are. We recorded 14 species in seaside habitats, but include only 4 that are among those most easily recognized. Most of the remainder are rare.

This species is an erect hairy perennial to 2 m tall. Leaflets narrow. Flow-

Lespedeza hirta

ers in dense racemes that are shorter to 1.5 times longer than leaves at the raceme bases. Calyx hairy, 7–12 mm long. Petals yellowish white with a purple spot, standard 8–12 mm long. Common. Between dunes, thin woods, disturbed areas; SC–sME. July–Oct.

L. hirta (L.) Hornem. has similar flowers, but standard 6–8 mm long and calyx 4–8 mm long. Leaflets are from widely oblong to suborbicular. Flower clusters numerous, varying from mostly axillary to mostly on distinct peduncles, always some peduncled. Occasional. Similar places; eTX–ME. July–Oct.

Trailing Lespedeza
Lespedeza repens (L.) Bart.

[86]
Perennial with several to many trailing stems. Hairs on stems few to many and upwardly appressed. Stipules 2–4.5 mm long. Flowers and fruits 1 to many, short-stalked, in short racemes on slim prominent peduncles. Rare. Dry places—thin woods, roadsides, grassy areas; SC–NY. June–Sept.

L. procumbens Michx. is quite similar but has spreading hairs on stems. Occasional. Similar places; VA–NH. June–Sept.

Narrow-leaved Vetch
Vicia angustifolia L.

[87]
The vetches in seaside habitats are vinelike and have tendril-bearing pinnately compound leaves. The style is bearded all around the top, like a bottle-brush. Vetches are easily confused with species of *Lathyrus* but in the latter the style is bearded along one side of the tip, like a toothbrush. Vetches, like Lespedezas, are an important source of food for animals. Seeds of some species may be poisonous.

This species is a glabrous or nearly so annual to 60 cm long. Leaves with 4–10 leaflets. Flowers 10–18 mm long, commonly in pairs, nearly sessile in upper leaf axils. Mature fruits blackish. Seeds subglobose. Common. Weedy—roadsides, fencerows, fields, grassy areas; eTX–ME. Mar.–Oct. Syn.: *V. sativa* L. ssp. *nigra* (L.) Ehrh.

Sand Vetch
Vicia acutifolia Ell.

[88]
Perennial to 1.2 m long, spreading or climbing. Leaflets 2–6, usually 4, linear to linear-oblong. Flowers small, 7–8 mm long, 4–10 on peduncles longer than leaves; corolla pale blue, sometimes faintly so. Pods 23–30 mm long, seeds 4–10. Common. Sandy moist to wet soils in thin woods or in open—margins of marshes and ponds, sloughs, ditches, roadsides; AL–sSC. Feb.–June.

Two other species also have small flowers, under 9 mm long, on long peduncles. *V. tetrasperma* (L.) Moench, an annual, has 4–10 narrow

Vicia ludoviciana

Vicia tetrasperma

leaflets; 1–2, rarely 4, flowers per peduncle; and fruits 10–15 mm long with 3–6 seeds. Occasional. Dry to wet places—margins of marshes and ponds, ditches, woods, waste places; MS–Nfld. Mar.–June. *V. ludoviciana* Nutt. has 6–12 linear-oblong to oval leaflets, 2–12 lavender-blue flowers per peduncle, fruits 20–30 cm long with 4–8 seeds. Occasional. Broadleaf or pine woods, behind stable dunes; eTX–AL. Mar.–May.

Smooth Vetch
Vicia dasycarpa Ten.

[89]
Stems glabrous or with a few appressed hairs. Leaflets 14–20. Flowers 12–17 mm long, 5–15 in dense racemes on peduncles longer than leaves. Calyx bulging at base, the pedicel appearing lateral. Occasional. Roadsides, fencerows, waste areas; NC–ME. May–Sept. Syn.: *V. villosa* ssp. *varia* (Host) Corb.

There are two similar species with similar habitats that may be encountered, as follows: *V. villosa* Roth with spreading hairs on stems, flowers 12–20 mm long and 10–40 per raceme. Common; VA–ME. May–Sept. *V. cracca* L. with flowers 9–13 mm long and 20–50 per raceme. The calyx is scarcely bulging at base; the pedicel appears basal. Occasional; MD–Nfld. May–Aug.

Vicia cracca

Vicia dasycarpa

Beach-pea
Lathyrus japonicus Willd.

[90]
Lathyrus species are much like vetches. To distinguish, see under *Vicia angustifolia*. Some species are poisonous if eaten.

Beach-pea has 4–12 leaflets; stipules nearly as large as leaflets, broadly ovate, symmetrical, with 2 basal lobes. Peduncles with 3–10 flowers. Common. Dunes, seashores, pond shores, roadsides, fields; NJ–Nfld. May–Sept. *L. maritimus* (L.) Bigel. of some manuals.

L. palustris L. is similar but the stipules are narrowly lanceolate to nearly ovate and have only 1 basal lobe. Rare. Wet meadows, depressions between dunes, marsh edges; NY–Nfld. May–July.

Butterfly-pea
Clitoria mariana L.

[91]
A trailing or occasionally twining perennial. Leaflets 3, entire. Calyx tube much longer than lobes. Standard 4–6 cm long. Stamens 10, united into a tube. Legume with a long stipe, flattened, about 6 mm wide, 3–6 cm long. Common. Well-drained soils—dry flats between and behind stable dunes, thin woods, grassy areas; eTX–NY. June–Aug.

Climbing Butterfly-pea
Centrosema virginianum (L.) Benth.

[92]
Twining perennial from a tough elongated root. Leaflets 3. Calyx tube shorter than the lobes and hidden by conspicuous bracts. Petals lasting 1 day. Standard petal 25–35 mm long, about twice as long as the wing and keel petals. Legume sessile, flattened, many-seeded, about 4 mm broad, 7–14 cm long, with a persistent style. Common. Well-drained places—thin woods or in open, including dunes; eTX–sNJ. Mar.–Sept. Syn.: *Bradburya virginiana* (L.) O. Ktze.

Cardinal-spear; Coral Bean
Erythrina herbacea L.

[93]
Perennial to 1.2 m tall. Leaves alternate with 3 leaflets that are hastate to widely deltoid and occasionally prickly beneath. Stipules are curved spines. Inflorescence of one or more terminal spikelike racemes. Calyx red, tubular. Corolla scarlet, the standard to 53 mm long, folded so that the entire flower appears long and narrow. Wing and keel petals 13 mm or less long. Fruit to 21 cm long, constricted between the few to many brilliantly scarlet seeds that often remain attached long after the pod splits open. Occasional. Thin woods or in the open, dunes, sandy soils; TX–sNC. Apr.–July.

In the warmer parts of the range of the species, aboveground stems may survive over a number of winters. This has led some people to believe there is another species, *E. arborea* (Chapm.) Small, but this is not supported by present data.

Milk-pea
Galactia volubilis (L.) Britt.

[94]
Herbaceous perennial twining vine. Stem hairs retrorse to spreading. Leaflets 3, entire. Flowers 10–14 mm long, on axillary racemes, in separated nodes and not congested near the tip of the raceme, each flower with 2 minute bracts at or near the top of the pedicel, as on all *Galactia* species. Longest of the inflorescences 5–55 cm long. Calyx lobes 4 by fusion of the upper 2. Petals pink to purplish, the keel petals nearly straight. Fruit a legume with 10–13 seeds, including aborted ones. Common. Climbing vegetation at edge of marshes, woods, and thickets; between stable dunes, thin woods, open areas; LA–sVA. June–Sept. Syn.: *G. macreei* M. A. Curtis.

G. regularis (L.) B.S.P. (*G. mississippiensis* Vail) is quite similar but has smaller flowers, 7–9 mm long, and at tip of raceme they are congested. Occasional. Similar places; eTX–sVA. May–Sept. In *G. floridana* T. & G., hairs on the stem are fine, close, mostly spreading, the longest 0.7 mm or longer. Petals falling as they wither or soon thereafter. Occasional. Usually dry places, sterile sand, scrub areas, stable dunes; MS–wFL. June–Sept.

Galactia
Galactia elliottii Nutt.

[95]
Twining, climbing perennial often with a long horizontal rootstock and some tuberous roots. Leaflets 7 or 9. A pair of very small bracts at or near top of each pedicel. Legumes, 3–5 cm long, 6–13 mm wide, densely covered with short appressed hairs. Occasional. Sandy areas—thin pinelands, thin broadleaf woods, between stable dunes, thin scrub areas, grassy areas; lower wFL–sSC. May–Sept.

Climbing Rhynchosia
Rhynchosia minima (L.) DC.

[96]
Perennial vine with 3-foliate leaves, blades entire. Flowers in axillary racemes 45–160 mm long. Sepals united. Filaments filiform, 9 united and 1 separate. Anthers all alike. Legume 9–20 mm long. Seeds eaten by animals, especially birds. Occasional. Thin pinelands, hammocks, stable dunes, thin scrub; TX–GA. Apr.–Dec.

Wild Bean
Strophostyles helvola (L.) Ell.

[97]
Trailing or twining annual. Leaves with 3 entire leaflets. Flowers few in a dense cluster on an axillary peduncle 15–30 cm long. Small bracts below each flower acute, lanceolate, and about as long as calyx tube. Standard 1–2 cm broad, tip of keel petals and style strongly curved back toward their bases. Legume 3.5–10 cm long. Common. Sandy soils—beaches, low

Strophostyles helvola

Strophostyles helvola

dunes, edge of marshes; eTX–sME. June–Sept.
　　S. umbellata (Muhl. ex Willd.) Britt. is similar but is a perennial and the bracts below each flower are blunt and about half as long as calyx. Occasional. Fields, thin woods, sandy soils, dunes; eTX–NY. June–Sept.

Vigna; Savi
Vigna luteola (Jacq.) Benth.

[98]
Perennial vine much like *Strophostyles* vegetatively but the corollas are yellow and the keel petals not so strongly curved. Legume 3–7 cm long. Occasional. Moist places—lagoon shores, beaches, dunes, thin woods, roadsides; TX–sNC. Mar.–Nov.

GERANIACEAE
Geranium Family

Carolina Cranesbill
Geranium carolinianum L.

[99]
Annual to 60 cm tall with deeply dissected but simple leaves. Pedicels about as long as calyx. Sepals with subulate tips 1 mm long or longer. Petals 4–6 mm long. Stamens 10. Mature fruits with a prominent beak developed from the style, the 5 carpels with 1 seed each and with spreading hairs over 0.6 mm long. Common. Dry places—thin woods, roadsides, fields, waste places, between stable

dunes; TX–MA. Mar.–July.

Two similar species of limited distribution may be encountered. *G. texanum* (Trel.) Heller has short ascending to appressed hairs on the carpels. Rare. Sandy soils in thin woods or open; TX–LA. Mar.–Apr. *G. robertianum* L. may be recognized by leaves completely divided into 3–5 separate pinnately cleft segments. Rare. Moist places—pond margins, woods, swamps; gravelly shores; NY–Nfld. May–Sept.

OXALIDACEAE
Wood-sorrel Family

Yellow Wood-sorrel
Oxalis stricta L.

[100]

In *Oxalis* species leaves are palmately compound, the three leaflets obcordate. Sepals and petals 5. Stamens 10, of 2 lengths. Carpels 5. Fruit a cylindrical capsule, several to many-seeded. Plants are usually difficult to name to species. We have recorded 6 species in seaside habitats. Only 2 are common.

This species is a perennial to 50 cm tall, with slender rhizomes. Stems decumbent to erect, with small whitish appressed hairs. Leaflets 1–2 cm broad. Stipules absent. Flowers 7–11 mm long in peduncled 1- to 4-flowered clusters. Capsules with short scattered spreading hairs. Common. Stable dunes and interdunes, thin woods, fields, waste places; TX–NS. May–frost.

O. dillenii Jacq. is similar but rhizomes are short or absent and stipules are present and oblong. Common in the south but rare north of VA. Similar places; TX–NS. Mar.–frost.

LINACEAE
Flax Family

Wild Flax
Linum medium (Planch.) Britt.

[101]

Perennial to 60 cm tall with unbranched terete stems, 20–40 leaves. Lower leaves mostly alternate, elliptic to oval or elliptic-obovate, 1–4 mm wide, 1–2 cm long, without obvious veins. Branches of inflorescence stiffly ascending. Sepals, petals, and stamens 5. Inner sepals densely glandular-serrate on margins, persisting at maturity of the subglobose capsule. Capsule 2.0–2.5 mm broad, 1.6–2.3 mm long, 5-carpeled but splitting into 10 half-carpels at maturity. Seeds 10. Common. Depressions, marshes, bogs, fields, meadows, swales; eTX–sME. June–Oct.

L. virginianum L. is quite similar, but leaves are 2–7 mm wide and inner sepals are scattered glandular-serrate on margins. Fruit 1.3–1.8 mm long. Common. Similar places; VA–MA. June–Oct. *L. striatum* Walt. also has few glands on the inner sepals but the stem has 3 narrow wings extending from each leaf to beyond the leaf below. Occasional. Damp sands, swamps, bogs, freshwater marshes, edge of ponds, swales, ditches; MS–wFL; SC–sMA. June–July.

Linum medium

POLYGALACEAE
Milkwort Family

Large-flowered Polygala
Polygala grandiflora Walt.

[102]
Polygalas have flowers in racemes or spikes. Flowers are perfect, irregular, with 3 small sepals and 2 petal-like ones called wings. The corolla resembles those of some of the bean family, as do the stamens, which are united by their filaments and to the petals. Polygalas are easily separated from legumes by the 2-carpelled ovary and fruit. Seeds 2.

This species is a perennial to 50 cm tall. Stem usually unbranched, 1 to several, with appressed to spreading hairs. Leaves alternate, oblanceolate to linear-oblanceolate, 1.5–5 cm long. Flowers 6–7 mm long, largest sepals as wide as long. Corolla not fringed. Common. Sandy soils—thin pinelands, roadsides, pine-palmetto; eLA–sSC. Apr.–Sept.

Candyweed; Polygala
Polygala lutea L.

[103]
Glabrous biennial or occasionally perennial to 42 cm tall. Stems usually unbranched, 1 to several from base, with appressed to spreading hairs. Leaves alternate, lower ones spatulate, persisting; upper ones oblanceolate to linear-oblanceolate, 1.5–5 cm long. Racemes orange, drying to yellow, 1 on each stem. Sepals under 1 mm long. Pedicels winged. Occasional. Moist soils—thin pinelands, grassy areas, roadsides; MS–sSC. Feb.–Nov.

Plants to only 13 cm tall, with leaves all basal or nearly so, and lemon-yellow racemes are *P. nana* (Michx.) DC. Rare. Sandy soils—thin pinelands, roadsides, edge of boggy areas; MS–wFL. Feb.–June.

Slender Polygala
Polygala incarnata L.

[104]
Very slender glaucous annual to 60 cm tall, with scattered alternate linear leaves 5–15 mm long that frequently fall off early. Stems simple or with a few erect branches. Flowers nearly sessile. Sepals 2.5–3 mm long, pink. Corolla reddish pink, 6–7 mm long, with a reddish tube. In all other species the corolla is less than twice as long as longest sepal. Common in south to rare in north. Thin pinelands, meadows, bogs, swales, roadsides; eTX–NY. June–Nov.

EUPHORBIACEAE
Spurge Family

Croton
Croton glandulosus L.

[105]
Annual to 60 cm tall, bearing fine stellate hairs. Roots with spicy odor. Juice not milky. Leaf blades crenate-serrate, oblong to lanceolate, 2–9 cm long. Male and female flowers on same plant, male flowers with 4-parted calyx and 4 white petals. Fruit a capsule with 1 seed in each of the 3 carpels. Common. Dry sand and loam soils, thin woods, roadsides, meadows, between stable dunes; eTX–VA. May–Oct.

Three-seeded Mercury
Acalypha gracilens Gray

[106]
Erect annual to 80 cm tall, simple or little branched from near base. Stems bearing small incurved hairs. Leaf margins entire or nearly so. Male and female flowers on same short spike, which is axillary and has a leafy bract at the base. Bracts with 9–15 ovate to deltoid lobes 2 mm or less long. Fruit a 3-seeded capsule. Seeds reddish to black. Occasional. Broadleaf and pine woods, fields, waste places, mead-

ows, edge of brackish marshes; eTX–sME. June–frost.

Two other species are similar. In *A. rhomboidea* Raf. the bracts have only 5–7 lobes, the longest 3 mm or longer. Occasional. Similar places; LA; SC–NS. June–frost. In *A. virginica* L. stems have straight spreading hairs, especially on the upper part. Rare. Thin woods, disturbed areas, roadsides; VA–MA. Aug.–frost.

Bull-nettle, Mala Mujer
Cnidoscolus stimulosus (Michx.) Engelm. & Gray

[107]
Erect perennial herb to 1 m tall from a large root. Sepals white, petals none. Tube of male flowers glabrous, under 1 cm long. Fruits with 3 hard seeds 8–9 mm long. Stems, leaves, and female flowers bear long stiff sharp hairs with a caustic irritant that on contact produces painful irritation and in some people a severe reaction. Common. Sandy, often disturbed soils, dunes, meadows, thin woods; eLA–sVA. Mar.–Sept. Syn.: *Jatropha stimulosus* Michx.

C. texanus (Muell. Arg.) Small is quite similar but the tube of the male flowers is spiny and 1–2 cm long, and the seeds 12–15 mm long. Occasional. Sandy areas, dunes, grasslands; TX–wLA. Mar.–Nov.

Seaside Spurge
Chamaesyce polygonifolia (L.) Small

[108]
Seaside *Chamaesyce* species are prostrate to ascending annuals with milky juice and are usually much-branched from base upward. Leaves are opposite, blades asymmetric at base. Flowers are unisexual. Both sexes are borne together in cups (cyathia) usually with 4 or 5 petal-like lobes, the whole resembling a flower. Male flowers consist of a single stamen, female flowers of a single 3-carpelled pistil. Fruits contain 3 seeds, 1 in each cavity, and split apart. Plant juices may cause skin irritation, and plants when eaten may cause severe poisoning.

This species has glabrous prostrate to ascending stems. Leaves entire, usually mucronate, about 3 times as long as wide. Fruits 3–3.5 mm long. Common. Dunes, sandy flats, upper beaches; LA–NBr. May–frost. Syn.: *Euphorbia polygonifolia* L.

C. bombensis (Jacq.) Dug. is quite similar but fruits about 2 mm long. Common. Dunes, sandy barrens; TX–sVA. May–frost. Syn.: *Euphorbia ammannioides* H.B.K. In *C. cordifolia* (Ell.) Small leaves are about twice as long as wide or shorter, otherwise closely resembling the above 2 species. Rare. Sandy soils out of dune areas, thin woods or in the open; TX–wFL. May–frost.

Prostrate Spurge
Chamaesyce maculata (L.) Small

[109]
Stems prostrate, finely hairy on all sides. Leaves with very small teeth near tip, peduncles under 6 mm long. Fruit hairy. Common. Along paths, crevices and sides of sidewalks and roads, yards, flower borders, waste places; TX–ME. Apr.–frost. Syn.: *Euphorbia supina* Raf.

C. prostrata (Ait.) Small is similar but the stem is hairy in a line on 1 side only. Rare. Similar places; TX–wFL. May–frost.

Eyebane; Wartweed
Chamaesyce nutans (Lag.) Small

[110]
Erect to spreading branched plants to 80 cm tall. Stems glabrous or hairy on 1 or 2 sides. Leaves oblique at base, finely serrate, often with a purple spot at middle. Cyathia single or a few together, peduncles to 5 mm long. Fruit glabrous. Seeds obtusely angled, the faces with 5–9 faint ripples on surface. Common. Waste places, open areas, fields, meadows; TX–NBr. May–Nov. *Euphorbia maculata* L. of most manuals.

Hairy Spurge
Chamaesyce hirta (L.) Millsp.

[111]
Stems erect to decumbent, 2–60 cm long, with conspicuous hairs. Leaves strongly asymmetrical at base. Cyathia very small, numerous, borne in dense conspicuous clusters on peduncles about 1 cm long or longer. Capsules finely hairy, sharply 3-angled, 1–1.2 mm long. Occasional. Fields, pastures, roadsides, beside paths, around buildings, dock areas; LA–GA. Rare in TX and SC. Apr.–frost.

Cypress Spurge
Euphorbia cyparissias L.

[112]
Reproductive structures of seaside *Euphorbia* species are like those of *Chamaesyce*. The genera differ vegetatively in that seaside *Euphorbia* species are perennial, have a prominent main stem, sometimes several from the rootstock, and alternate leaves that are symmetrical.

Plants to 70 cm tall from extensive and branching thin rootstocks. Mid-stem leaves entire, alternate, 1–2 mm wide, 1–3 cm long, sessile. Floral leaves entire, 4–6 mm long. Cyathia 3 mm high, their glands waxy yellow. Occasional. Waste areas, roadsides, around residences, sandy flats between and behind dunes; NY–NS. Apr.–Aug.

E. esula L. has similar floral bracts but is 10–13 mm long and the mid-stem leaves are 2–10 mm wide. Stems arise from deep roots. Rare. Similar places; NJ–NS. May–Aug.

BALSAMINACEAE
Touch-me-not Family

Jewelweed; Touch-me-not
Impatiens capensis Meerb.

[113]
A succulent annual to 2 m tall, heavily glaucous, repelling water, but drops sometimes stand on horizontal areas and appear like jewels in reflected light. Some flowers conspicuous, orange-yellow spotted with brown, on slim usually drooping pedicels. Sepals 3, colorful, one of them forming a prominent sac with a curled spur at the base. Other flowers very small, these primarily producing the fruits. Fruits usually drooping, 5-carpelled, green, and coiling elastically into 5 sections when mature, often ejecting the seeds considerable distances. Emetic and poisonous when eaten. Juice of plant has been used to treat irritation from poison ivy. Common. Wet, freshwater habitats, in woods or in the open; VA–Nfld. June–frost. Syn.: *I. biflora* Walt.

MALVACEAE
Mallow Family

Sida
Sida rhombifolia L.

[114]
Annual or biennial to 1.2 m tall, the stem tough. Flowers solitary in leaf axils, on peduncles several times longer than petioles. Stamens united by filaments into a tube, arranged in 5 bundles near the top. Fruit a capsule with 1 seed in each carpel. Common. Roadsides, waste places, thin woods; eTX–sNC. Annual and biennial plants in June–Oct.; biennials again in Apr.–May.

Two similar species may be encountered in seaside habitats. Both have peduncles no longer than the petioles. *S. spinosa* L. can be identified by presence of hard stipular spines.

Sida acuta

Rare. Similar places, usually moist; eTX–AL. June–frost. *S. acuta* Burm. f. has abruptly tapering leaf bases and slightly larger flowers. Occasional. Thin woods, behind stable dunes; waste places; MS–sSC. June–Oct. Syn.: *S. carpinifolia* L. f.

Swamp Rose-mallow
Hibiscus moscheutos L.

[115]
H.m. var. *moscheutos* is a perennial to 2 m tall. Leaves soft hairy beneath, glabrous above, unlobed or the lower ones shallowly lobed. Petals white, or less often pink, but always with a purple to reddish center; 10–12 cm long. Fruit a glabrous 5-carpelled capsule. Occasional. Open swamps, fresh and brackish marshes, sloughs; eTX–CT. May–Sept.

In the similar *H. m.* var. *palustris* (L.) Clausen the petals are pink to white but lack a central crimson-purple spot at center. Occasional. Salt, brackish, or fresh marshes; sloughs; VA–sMA. June–Sept.

H. lasiocarpus Cav. is also similar but the leaves are grayish stellate hairy above and the seed pod is hairy. Rare. Open swamps and marshes; eTX–AL. June–Oct. Syn.: *H. incanus* Wendl. Huge flowers, petals 12–14 cm long, identify *H. grandiflorus* Michx. Occasional. Freshwater marshes, ponds, sloughs; eLA–sGA. June–Sept.

Seashore Mallow
Kosteletzkya virginica (L.) Presl ex Gray

[116]
Perennial to 1.5 m tall. Leaves densely hairy, cordate-orbicular to lanceolate, with or without hastate lobes at base. Base of calyx tube with 8–10 bracts 1–3 mm wide. Stigmas, styles, and carpels 5. Fruit prominently 5-angled, flattened at top, covered with hairs 1.5–2 mm long. Seeds 5, smooth. Common. Salt, brackish, or fresh marshes; thin swampy woods; sloughs; TX–NY. June–Oct.

STERCULIACEAE
Cacao Family

Chocolate-weed
Melochia corchorifolia L.

[117]
Annual with tough stems to 1.5 m high. Flowers mostly in a dense terminal head, flower parts 5 each. Calyx partly united, persistent, with 3 or more narrow bracts at base. Petals short-clawed, pink to purple or white. Filaments partly united. Styles separate. Capsules smooth, 4–4.5 mm long. Seeds smooth, 2, or 1 by abortion, in each of the 5 carpels. Occasional. Weedy—stable dune areas, roadsides, thin woods or scrub, eTX–sSC. Aug.–Oct.

In *M. villosa* (Mill.) Fawc. & Rendle the flowers are sessile in leaf axils and not in a dense terminal head. Styles partly united. Rare. Thin woods and scrub; FL–sGA. Aug.–Oct. Syn.: *M. hirsuta* Cav.

Melochia corchorifolia

HYPERICACEAE
St. John's-wort Family

Common St. Peter's-wort
Hypericum perforatum L.

[118]
Hypericums are woody or herbaceous. Leaves entire, opposite; blades

ordinarily with translucent internal glands visible with transmitted light, or sometimes glands dark. A magnifying lens may be needed to see them. Sepals 4–5, separate. Petals 4–5, separate, yellow to orange-yellow. Ovary superior, hypanthium absent. Fruit a 1- or 3- to 5-celled, 3- to 5-carpelled capsule with many small seeds.

Hypericum punctatum

Hypericum perforatum

This species is a much-branched perennial to 90 cm tall. Leaves sessile, spreading, linear-oblong, the mid-stem leaves 2–4 cm long with transparent dots. Petals 5, often black-dotted, 7–10 mm long. Stamens many, united at base into 3–5 tufts. Carpels 3, styles 3–5 mm long. Seeds roughened by a coarse network of ridges. Common. Weedy in open places—meadows, between stable dunes, marsh edges, protected beaches, roadsides, fields; nVA–Nfld. June–Sept.

H. punctatum Lam. is similar but is sparingly branched and the leaves are dark-dotted. Petals only 4–7.5 mm long. Common. Similar places; VA–ME. June–Sept.

Pineweed; Orange-grass
Hypericum gentianoides (L.) B.S.P.

[119]
Annual to 50 cm tall. Leaves subulate, usually less than 5 mm long, appressed. Flowers sessile or nearly so. Stamens in clusters, fastened together at their bases. Mature capsules to 5.5 mm long, over twice as long as the sepals. Common. Usually in poor soils—stable dunes, sandy flats, thin maritime woods, roadsides, dry meadows; eTX–sME. July–Oct. Syn.: *Sarothra gentianoides* L.

Hypericum
Hypericum mutilum L.

[120]
Perennial to 80 cm tall, usually much-branched. Midstem leaves ovate-oblong to elliptic, rounded to blunt at apex. Sepals broadest near middle, acute, about as long as fruit. Stamens 5–15. Style to 1 mm long. Fruit widest at middle. Common. Wet places—freshwater marshes, swamps, bogs, pond edges, ditches; eTX–NS. June–Oct.

H. boreale (Britt.) Bickn. is similar but usually smaller; the stems are often decumbent, sepals obtuse and much shorter than fruit. Common. Similar places; MD–Nfld. July–Oct. *H. gymnanthum* Engelm. & Gray also has

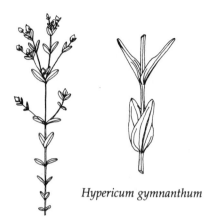

Hypericum gymnanthum

Hypericum boreale

Hypericum mutilum

Perennial to 80 cm tall with cordate-clasping leaves. Sepals 5, 3–8 mm long. Petals 5, pink, 7–10 mm long. Styles 3, 0.5–3 mm long. Rare in south to common in north. Bogs, swamps, freshwater marshes, pond margins; eTX–Nfld. July–Sept. Syn.: *Hypericum virginicum* L.

CISTACEAE
Rock-rose Family

Sun-rose; Frostweed
Helianthemum carolinianum (Walt.) Michx.

[122]
Helianthemum species have distinctive showy flowers. Petals 5, yellow. Stamens many, radiating. Style short, the 3 stigmas forming a conspicuous head. The many ovules and seeds, which are small, are fastened to 3 parietal placentas in the 1-celled 3-carpelled ovary.

In this species leaves are mostly basal, hairs on underside not hiding epidermis. Flowers few, not scattered. Petals 8–20 mm long. Occasional. Dry sandy soils—thin woods, thin scrub, stable dunes and interdunes, meadows; eTX–sNC. Mar.–May.

Three other species with similar flowers are occasional along the coast

few stamens but is little-branched, midstem leaves taper to a point, and sepals and fruit are widest below the middle. Occasional. Similar places; eTX–MD. June–Oct.

Marsh St. John's-wort
Triadenum virginicum (L.) Raf.

[121]
Triadenum species are much like *Hypericum* spp. but may be recognized by flesh-colored to purplish or pinkish petals and 3 small orange glands alternating with 3 clusters of 3 stamens each.

Helianthemum canadense

in dry sandy soils. Stems often in clusters. Leaves are scattered along the stem and have fine dense hairs below. Flowers in clusters. Petals 10–15 mm long. In *H. canadense* (L.) Michx. calyx of petal-bearing flowers has spreading hairs and seeds are papillose. Similar places; nNC–ME. Apr.–July. In *H. georgianum* Chapm. calyx is 4–6 mm long and hairs are short and appressed. Seeds finely pitted. Similar places and maritime woods; eTX–NC. Mar.–May. *H. arenicola* Chapm. is most like the latter species but the calyx is 6–8 mm long. Coastal dunes and sands; MS–wFL. Mar.–May.

Pin-weed
Lechea racemulosa Michx.

[123]
Pin-weeds are conspicuous for abundant, though small, flowers and fruits. Sepals, petals, and fruits are about 2 mm long or less. Sepals of 2 sizes, the outer ones much narrower than inner ones. Petals 3, reddish. Stigmas sessile, with narrow thread-like divisions. Pin-weeds are perennials to 90 cm tall with first stems erect, later with basal rosettes of prostrate leafy stems. One or more species occur in seaside habitats from TX to NBr, in dry places in woods or in the open, such as established dune areas. They fruit from June (in south) to frost. All are hairy, some finely so or nearly glabrous. Leaves vary from linear to elliptic, oblong, or oblanceolate. Minute structures and other factors make most species difficult to recognize.

This species is finely appressed-hairy, fruiting stems to 45 cm long. Leaves glabrous above. External sepals shorter than internal ones. Capsule and calyx together 1.5 times or more as long as wide. The photograph shows only a few of the many fruit-bearing branches. In various species fruits vary to nearly globose in shape and may be more abundant and denser. Occasional; VA–NY. June–July.

VIOLACEAE
Violet Family

Lance-leaved Violet
Viola lanceolata L.

[124]
Seaside violets are perennials with irregular corollas, the lower petal spurred at base. Flowers arise directly from a rhizome. Species are often hard to identify. Hybrids are frequent.

This species has leaves 3.5–15 times as long as broad. Petals white with purple veins, the 2 lateral ones with a tuft of hairs on the inside. Occasional. Wet sandy soils—shores of ponds, meadows, bogs, freshwater marshes, swamps; eTX–NS. Mar.–June.

Sometimes divided into subspecies: *lanceolata*, which is in the photograph, from VA to NS, with leaves 3.5–5 times as long as broad; and *vittata* (Greene) Russell from eTX to sNC, with leaves 6–15 times as long as broad.

Most of the other 7 violets recorded for seaside habitats are quite rare. Of these, *V. septemloba* LeConte from MS to sNC is the most likely to be encountered. The corolla is blue-violet with a broad, nearly white center; rootstock thick and fleshy; leaves glabrous, about as long as wide, lobed except sometimes the first 1 or 2. Thin pine woods; MS–SC. Mar.–Apr.

Viola septemloba

Viola septemloba

PASSIFLORACEAE
Passion-flower Family

Passion-flower
Passiflora incarnata L.

[125]
Passion-flowers are tendril-bearing perennial vines with distinctive flowers that possess a many-segmented crown above the perianth. The 5 stamens are united by their filaments into a tube surrounding the stalk of the ovary, which is visible above the tube. The stigmas are on the underside of the tips of 3 spreading styles. Fruit is a many-seeded berry with 3 parietal placentas.

This species has spectacular flowers 6–8 cm wide. Leaves are lobed and the lobes toothed. Fruit 4–7 cm long, fleshy. Often called maypops because of the fruits. Occasional. Fields, roadsides, fencerows, thin woods, thin scrub; eTX–MD. June–Oct.

P. lutea L. has entire obtusely 3-lobed leaves. Flowers about 2–3 cm wide. Petals yellowish green. Crown greenish. Fruit 8–15 mm long. Occasional. Thin woods, thickets, depressions between stable dunes, slough edges, often in moist soils; eTX–VA. June–Oct.

LYTHRACEAE
Loosestrife Family

Toothcup
Rotala ramosior (L.) Koehne

[126]
Annual to 30 cm tall. Stems glabrous. Leaves opposite, simple, entire, base tapered, 1–4 cm long, linear-lanceolate to oblanceolate. Flowers one in each leaf axil. Calyx lobes with acute tips. Petals 4, separate, white or pink. Style about 0.5 mm long. Fruit a capsule borne free within a subglobose hypanthium. Occasional. Wet places in open—freshwater and brackish marshes, ditches, margins of ponds, sloughs; eTX–wFL. Apr.–Nov.

Two species of *Ammannia* are similar, but there are 2 to several flowers in leaf axils, the petals are purple to pink, and leaf bases usually auricled to cordate. In *A. latifolia* L. the style is 0.5 mm long and tip of the calyx lobes obtuse and often mucronate. Common. Similar places; TX–NJ. July–Oct. Syn.: *A. teres* Raf. In *A. coccinea* Rottb. the style is 1.5–3 mm long and tip of the calyx lobes acute. Rare. Similar places; TX–wFL. July–Oct.

Spiked Loosestrife
Lythrum salicaria L.

[127]
Stout erect perennial to 1.7 m tall, glabrous or hairy. Midstem leaves opposite, lanceolate to nearly linear, to 10 cm long, obtuse to cordate at base. Flowers in spikelike panicles. Hypanthium present, longer than wide. Petals 7–10 mm long, of equal size. Stamens twice as many as petals. Often an undesirable weed. Common locally. Flats between and behind dunes, marshes, pond shores, ditches, swales; NJ–Nfld. June–Sept.

Narrow-leaved Loosestrife
Lythrum lineare L.

[128]
Erect slender perennial to 1.2 m tall, with many strongly ascending flower-bearing branches. Upper part of stem quadrangular. Leaves all opposite, linear to linear-lanceolate, with acute tips and tapered bases. Flowers usually solitary in the axils of small leaves or bracts. Hypanthium tapering from top to bottom. Petals 1.5–3 mm long. Stamens same number as petals. Common. Freshwater, brackish, or saline marshes, swales, meadows, edge of sloughs and ponds; LA–NY. July–Oct.

L. alatum Pursh var. *lanceolatum* (Ell.) T. & G. ex Roth is similar but leaves from midstem upward are mostly alternate and petals 3–6 mm long. Occasional. Freshwater marshes, swamps, ditches, pond shores; eTX–sNC. May–Oct.

Lythrum lineare

MELASTOMATACEAE
Meadow-beauty Family

Smooth Meadow-beauty
Rhexia alifanus Walt.

[129]
This is the only genus in the United States of this large mostly tropical family of about 4000 species, three-fourths of which are in the Americas. The 13 species of *Rhexia* are confined to the United States except one that also occurs in the West Indies. *Rhexia* species are perennials with opposite leaves, an urn-shaped hypanthium that tightly encloses the 4-carpelled ovary and continues above it, asymmetrical petals, and 8 stamens with anthers opening by pores at the apex.

Smooth Meadow-beauty is sparingly branched, to 1.2 m tall. It has a completely hairless stem, prominently curved anthers, and a glandular-hairy hypanthium. The leaves are mostly lanceolate-ovate to elliptic or lanceolate, to 75 mm long at midstem

where they are the largest. Petals 25 mm long, spreading. Common. Sandy and peaty soils—low pinelands, bogs, savannas, freshwater marshes, between stable dunes; eTX–sNC. May–Sept.

Meadow-beauty
Rhexia nashii Small

[130]
Plants to 1.5 m tall, often forming extensive colonies from shallow elongate rhizomes. Stems hairy, 4-angled and thus 4-sided with 1 pair of opposing sides much narrower than the other pair. Midstem leaves mostly lanceolate to narrowly ovate or elliptic. Hypanthium 10–18 mm long. Petals 20–27 mm long, usually dull lavender. Anthers curved. Common. Acid sandy or peaty swamps, bogs, roadside ditches, swales, thin flatwoods; MS–sVA. May–Oct.

Of other similar species, 2 have unequal faces at midstem. In *R. mariana* L. var. *mariana* the petals are white to pale lavender, leaves linear-filiform to ovate or ovate-lanceolate. Hypanthium 6–10 mm long. Common. Similar places; eTX–NY. May–Oct. In *R. cubensis* Griseb. the petals are bright rose-lavender and midstem leaves mostly linear, oblong, or spatulate. Occasional. Similar places; MS–sNC. May–Oct.

Rhexia cubensis   *Rhexia nashii*

Common Meadow-beauty
Rhexia virginica L.

[131]
Stems to 1 m tall, often spongy-thickened near the base, with conspicuous wings on the 4 angles, the 4 faces equal or nearly so, glandular hairy or with only a few hairs at nodes. Hypanthium 7–10 mm long, neck shorter than body. Petals bright lavender-rose. Common. Wet places—thin pinelands, pond shores, ditches, bogs; MS–ME. June–Oct.

The 4 faces of the stem are also equal or nearly so in *R. mariana* L. var. *ventricosa* (Fern. & Grisb.) Kral & Bostick but the stem angles are wingless to quite narrowly winged. The hypanthium is 10–12 cm long and the neck about as long as body. Rare. Similar places; nSC–sNJ. June–Oct.

ONAGRACEAE
Evening-primrose Family

Seed-box
Ludwigia alternifolia L.

[132]
Members of this family in the United States are all herbs with separate petals (when present), an inferior ovary, hypanthium frequently prolonged beyond the ovary, and fruits usually many-seeded capsules.

In *Ludwigia* the hypanthium is not prolonged beyond the ovary. We recorded 20 species of *Ludwigia* in seaside habitats. Eleven of those most likely to be encountered are included here.

This species is an erect annual to 1.2 m tall, roots often with swollen sections. Leaves alternate, petioled, the blades tapering to base. Pedicels 3–5 mm long. Sepals, petals, and stamens 4. Ovary inferior. Capsule 4-angled or winged, cubical, about 4–6 mm wide. Common. Moist to wet places—thin woods, swamps, freshwater or brackish marshes, ditches, swales, sloughs, pond margins; eTX–MA. May–Oct.

L. alata Ell. has the same general aspect but is a perennial to only 80 cm tall, leaves nearly sessile, flowers sessile, petals missing, and capsule 3–4 mm wide. Leafy stolons are formed at base of plant in the fall. Occasional. Similar places; MS–sVA. June–Sept.

Seed-box; Primrose-willow
Ludwigia peruviana (L.) Hara

[133]
Conspicuous much-branched mostly erect perennial to 4 m tall, thinly to densely hairy. Leaves alternate, short-petioled, the blades 8–15 cm long, ovate to lanceolate or oblong-lanceolate. Petals 4 or 5, 2–3 cm long. Stamens 8 or 10. Ovary inferior. Fruit an obconic-cylindrical unwinged but 4- to 6-angled capsule with many seeds. Rare. Marshes, ponds, sloughs, ditches, swamps; LA–GA. June–frost.

In *L. uruguayensis* (Camb.) Hara the flowers are about the same size and appearance but the stems are creeping or floating, leaves often narrower, and the fruits are cylindrical and 10–15 cm long. Occasional. Similar places and mud flats; eTX–sNC. May–Oct.

Trailing Ludwigia
Ludwigia palustris (L.) Ell.

[134]
Glabrous prostrate perennial, often matted. Leaves opposite. Flowers and fruits sessile or nearly so and solitary in the leaf axils. Sepals 4. Petals absent. Stamens 4. Ovary inferior. Fruit a capsule, 2–4 mm long, with 4 green longitudinal bands. Seeds glabrous. Common. Wet to moist places or in water, on mud, pond or lake margins, swamps, swales, sloughs, ditches; eTX–NS. May–frost.

L. repens Forst. is quite similar but occasionally finely hairy, capsule lacking the green bands and 5–9 mm long. Occasional. Similar places; eTX–NC. May–Oct. Syn.: *L. natans* Ell. In *L. brevipes* (Long) Eames the pedicels of fruits are about 4–5 mm long. Rare. Similar places; FL–sNJ. May–Oct.

Narrow-leaved Ludwigia
Ludwigia linearis Walt.

[135]
Erect perennial to 80 cm tall with alternate sessile linear leaves. Flowers sessile in leaf axils. Sepals, petals, and stamens 4. Sepals ovate-deltoid. Ovary inferior. Capsule obconic, 4-sided, over twice as long as broad. Occasional. Ditches, muddy shores, often in shallow water, thin woods, swales, sloughs, freshwater marshes; eTX–sNJ. June–Sept.

L. linifolia Poir. also has linear leaves but the capsules are cylindric and 8–9 mm long, and the sepals linear. Rare. Similar places; MS–sNC. June–Sept.

Slender Seed-box
Ludwigia virgata Michx.

[136]
Erect little-branched perennial to 1 m tall with locally swollen roots. Leaves sessile, linear to lanceolate or narrowly elliptic. Sepals, petals, and stamens 4. Sepals ovate, reflexed. Style 7–10 mm long. Capsules cubical, 4-

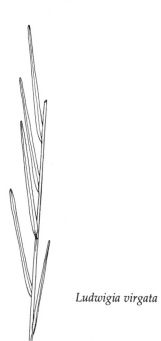

Ludwigia virgata

angled, 3.5–5 mm broad, on pedicels 10–15 mm long. Occasional. Swales, sloughs, bogs, savannas, ditches; MS–sVA. June–Sept.

Two other species are quite similar but the style is about 3 mm long or less. In *L. maritima* Harper the sepals are ovate and reflexed. Common. Similar places; eLA–NC. June–Sept. In *L. hirtella* Raf. the sepals are lanceolate to ovate-lanceolate and spreading to erect. Rare. Pinelands, ditches, bogs; AL–NC. June–Sept.

Beach Evening-primrose
Oenothera drummondii Hook.

[137]
Oenothera species have 4 separate petals and 8 stamens. Stigmas are deeply 4-lobed. The hypanthium is peculiar in that it not only surrounds and adheres to the sides of the ovary but is prolonged into a tube beyond the ovary. The prolonged portion falls off the maturing fruit.

This species is a spreading, decumbent, densely hairy perennial, sometimes shrubby in mild winters. Flowers in upper leaf axils. Petals 2–3.5 cm long, yellow. Capsule cylindric or nearly so, not winged. Seeds rounded, not angular. Occasional. Sandy beaches, dunes, sandy barrens, grasslands; TX–FL; SC. Mar.–Nov.

Dunes Evening-primrose
Oenothera humifusa Nutt.

[138]
Prostrate to decumbent perennial. Stems somewhat woody. Leaves acute, covered with fine densely matted grayish white hairs. Petals about 1 cm long. Capsule cylindrical or nearly so. Seeds rounded, not angled. Common. Dunes, dry sandy flats; LA–NY. May–frost.

Cut-leaved Oenothera
Oenothera laciniata Hill

[139]
Thinly haired annual or biennial. Stems usually decumbent. Leaves

nearly entire to irregularly cut. Flowers sessile in axils of the upper leaves, the inferior ovary appearing like a pedicel. Petals about 6 mm long, reddish brown as in photograph, or more commonly yellow, as in the preceding species, and 6–25 mm long. Capsule cylindrical, not winged. Seeds rounded, in distinct rows. Common. Open places—grasslands, seaside sands, brackish meadows, fields, waste places; eTX–ME. Mar.–July, into Oct. in north.

Evening-primrose
Oenothera biennis L.

[140]
Erect, hairy to nearly glabrous biennial to 2 m tall. Flowers in terminal clusters. Tips of calyx lobes in unexpanded buds are close together at end of the tapering bud, forming a tube. Capsule cylindrical. Seeds prismatic, horizontal in capsule. Common. Stable dune areas, grasslands, protected beaches, roadsides, around buildings, waste places; TX–Nfld. June–Oct.

O. parviflora L. is closely similar but is usually smaller, to 80 cm tall, and tips of calyx lobes in the unexpanded buds standing apart. Common. Edges of salt or fresh marshes, between stable dunes, protected beaches; NJ–Nfld. May–Sept. *O. heterophylla* Spach. is of a similar aspect. It grows to 1 m tall, has obovate petals 16–30 mm long. Seeds ascending in capsule, not sharply angled. Rare. Sandy places in open; eTX–wLA. June–Aug.

Sundrops
Oenothera fruticosa L.

[141]
Erect, hairy, usually branched perennial to 80 cm tall. Principal leaves elliptic to lanceolate or obovate, entire or nearly so. Flower buds erect. Part of hypanthium above ovary 7–15 mm long. Petals yellow, 1–2 cm long. Anthers 5–9 mm long. Stigmas beyond anthers when pollen is shedding. Capsule club-shaped, with appressed glandless hairs. Common. Meadows, thin woods, margins of freshwater and brackish marshes, stable dunes, roadsides; SC–NH. Apr.–Aug.

O. tetragona Roth is quite similar but the leaves are generally toothed and hairs on the capsule are predominantly gland-tipped and spreading. Occasional. Usually sandy soils of meadows, roadsides, thin woods, swales; SC–NS. May–Aug. Syn.: *O. fruticosa* ssp. *glauca* (Michx.) Straley. In the similar *O. perennis* L. the flowers are usually nodding, petals 5–10 mm long, and stamens the same length as the pistil. Common. Dry to moist places—thin woods, bogs, fields, meadows, gravelly slopes; DE–Nfld. May–Aug.

Showy Evening-primrose
Oenothera speciosa Nutt.

[142]
Erect to spreading perennial to 70 cm tall. Leaves linear to oblong-lanceolate, the wider ones irregularly dentate or narrowly lobed, especially near their bases. Flower buds nodding. Petals white to dark pink. Anthers 1.2–2 mm long. Fruit an ellipsoid to subglobose capsule. This species is hardy and drought-resistant and makes a showy ornamental. Occasional. Dry places—dune areas, meadows, fields, roadsides, waste places; TX–VA. Apr.–July.

Gaura
Gaura angustifolia Michx.

[143]
Slender annual to 1.8 m tall, usually not branched at base, principal leaves narrowly elliptic to lanceolate. Flowers and fruits sessile or nearly so. Hypanthium prolonged beyond the ovary as in *Oenothera*, this part about the same length as the ovary. Sepals 3–4, 2.5–8 mm long. Petals 3–4, white to pink, about 5 mm long. Fruit without a stipe, indehiscent, hairy, acutely 3-ribbed, 1- to 4-seeded, narrowly ellipsoid to ellipsoid. Common. Dune areas, meadows, grasslands, thin woods, roadsides; MS–NC. May–Oct.

In *P. palustris* L. the submerged leaves (i.e., those that develop under water) are similar to those of *P. pectinata* but the above-water leaves are merely serrate. Common. Freshwater marshes, pools, ditches, swamps, wet shores of ponds; eTX–NS. May–Oct.

Proserpinaca palustris

Gaura angustifolia

G. *filiformis* Small is a similar annual or biennial, with thinly hairy stems, well-branched above, to 4 m tall. Flowers opening near sunset. Sepals 4, 7–14 mm long. Petals 4, 7–13.6 mm long. Occasional. Shell deposits along bay shores, thin woods, open sandy areas; TX–MS. June–Dec.

HALORAGIDACEAE
Water-milfoil Family

Mermaid-weed
Proserpinaca pectinata Lam.

[144]
Uncommonly branched perennial to 50 cm tall. Leaves alternate, nearly all alike and quite deeply pinnately cut, with the midline of leaf 1 mm wide or less. Sepals 3. Petals none. Ovary inferior. Fruit hard, 3-angled, 3-seeded. Seeds yellow, oblong. Common. Peaty soils of swamps and savannas, ditches, pond margins, edge of brackish marshes; eTX–NS. May–Oct.

ARALIACEAE
Ginseng Family

Wild-sarsaparilla
Aralia nudicaulis L.

[145]
Perennial, distinctive with flowers and fruits in umbels on a long peduncle that arises, as does the single compound leaf, from the underground rhizome. Sepals 5, 0.5 mm or less long. Petals 5, white to greenish, 1–2 mm long. Ovary inferior. Fruit a 5-seeded berrylike drupe. Occasional. Moist to dry places—thin woods, under shrub thickets in stable dune areas; nVA–Nfld. May–July.

APIACEAE
Parsley Family

Seaside Pennywort
Hydrocotyle bonariensis Comm. ex Lam.

[146]
Members of this family are usually quite distinct from species of other families. All are herbs, leaves are alternate or basal and usually compound, flowers and fruits are in umbels although in some species not obviously, as flowers are sessile and thus in heads. Ovaries inferior. Fruits are dry and have 2 one-seeded sections that split apart at maturity, a schizocarp.

Seaside Pennywort is a succulent perennial, rooting from nodes of slender creeping stems. Leaves simple. Blades peltate, 3–10 cm wide, orbicular or nearly so. Inflorescence a compound umbel of many flowers, usually continuing for some time to develop new sections with numerous new flowers. Fruits and flowers thus can be present at the same time. Umbel branches may become 10 cm long. Calyx lobes lacking. Petals 5, white to light yellowish green. Fruit flattened, about 3 mm wide, composed of 2 seedlike sections. Common. Primary and stable dunes and swales, sandy marshes, swamps, sand flats, sloughs; eTX–NC. May–Nov.

Marsh Pennywort
Hydrocotyle umbellata L.

[147]
Similar to *H. bonariensis* but leaves only 1–4 cm wide and the umbel simple and with 15–50 flowers. Fruits deeply notched at base, 2–3 mm wide. Common. Pond shores, swales, slough edges, ditches, freshwater marshes; eTX–NS. Apr.–Nov.

H. verticillata Thunb. is quite similar vegetatively but leaves to 5 cm wide and the inflorescence a spike or raceme with flowers in whorls of 2–7, occasionally with 1 or a few branches. Common. Similar places; eTX–sMA.

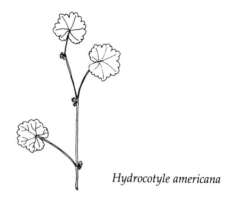

Hydrocotyle americana

Hydrocotyle verticillata

Apr.–Oct. *H. americana* L. has a similar aspect vegetatively but the orbicular leaf blades are cut to the petiole on one side. Flowers are often overlooked because they are in small umbels that are sessile or nearly so in leaf axils. Occasional. Moist places—in woods or in the open, meadows, swales, bogs, under shrubs; NY–Nfld. June–Sept.

Centella
Centella asiatica (L.) Urban

[148]
Perennial from thin rhizomes, resembling some *Hydrocotyle* species but petioles are attached to base of the cordate to truncate blades that are ovate to oblong and 2–10 cm long. Flowers 2–9, sessile or nearly so, in simple umbels. Peduncles 1–5 per node and 1–10 cm long, usually shorter than petioles. Common.

Sloughs, swales, meadows, ditches, pond shores, freshwater marsh edges; eTX–DE. May–Sept.

Chaerophyllum
Chaerophyllum tainturieri Hook.

[149]
Erect slender annual to 90 cm tall. Leaves much dissected, the ultimate segments 1–5 mm long and 0.75–2 mm wide, not linear. Petals white. Ovary and fruit glabrous. Fruits on short pedicels, narrowly ovate, 6–8 mm long, ribs broader than spaces between. Common. Roadsides, waste places, disturbed soils, around buildings, thin woods; eTX–VA. Mar.–June.

Water-hemlock
Cicuta mexicana Coult. & Rose

[150]
Cicuta are erect perennials from radiating tuberous roots resembling sweet potatoes. They are unusual in having hollow stems with a solid partition at nodes. Leaves alternate, those on lower half of stem at least twice-compound. Leaflets toothed and unusual in that the main lateral veins approach or end at the notches and not at tips of the teeth. Umbels compound. Petals white. Fruits glabrous. All parts are deadly poisonous to people and livestock when eaten.

This species is recognized by leaflets being mostly ovate to ovate-lanceolate and the 2 middle ribs on each side of the fruit having a narrow furrow between them. Common. Sloughs, pond margins, wet meadows, freshwater marshes, ditches; eTX–sNJ. June–Sept.

Two other species may be encountered. *C. maculata* L. has mostly lanceolate to lanceolate-oblong leaflets, and the 2 middle ribs of the fruits flush against each other. Common. Similar places; eTX–NS. May–Sept. The other, *C. bulbifera* L., has linear leaflets mostly less than 5 mm wide, and bulblets in axils of at least the upper leaves. Rare. Similar places; NY–Nfld. July–Sept.

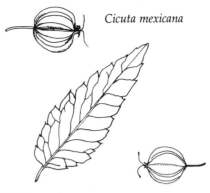

Cicuta mexicana

Cicuta mexicana

Cicuta maculata

Mock Bishop's-weed
Ptilimnium capillaceum (Michx.) Raf.

[151]
Glabrous annual to 80 cm tall. Stems few-branched. Petioles clasping stem. Leaves at least twice-compound, the filiform divisions not crowded. Petals white, rarely to pink. Fruits ovoid to suborbicular, in long-stalked once-compound umbels topping the adjacent leaf. Common. Brackish to freshwater marshes and swamps, sloughs, swales, pond margins, ditches; eTX–MA. May–Oct.

P. costatum (Ell.) Raf. is similar but more robust, to 1.5 m tall, and leaf divisions numerous, crowded, and appearing nearly whorled. Rare. Swamps, pond margins, sloughs; eTX–LA. June–Sept.

Ptilimnium capillaceum

Spermolepis divaricata

Ptilimnium costatum

Spermolepis divaricata (Walt.) Britt. is much like Mock Bishop's-weed but has very small warts on the fruits. Common. Dry to moist places—roadsides, yard borders, fields, swales; MS–NC. Apr.–May.

Lilaeopsis
Lilaeopsis chinensis (L.) Kuntze

[152]
At first glance vegetative stages of *Lilaeopsis* can easily be mistaken for small plants of some monocot. A close look reveals that the leaves (actually only petioles) are hollow and have cross partitions, which is unlike any seaside monocot with a similar aspect. Positive identification is afforded by the small-peduncled umbels of flowers or fruits. Calyx lobes are minute or absent. Petals 5, white. Fruits 2-carpelled, ribbed, about 2 mm long. Stems are slim, horizontal, and usually under soil surface, often forming dense mats.

In this species leaves are mostly 1–5 cm long and the peduncles as long or longer than the leaves. Common. In mud of brackish or salt marshes and tidal flats; LA–NS. Apr.–Sept.

In *L. carolinensis* Coult. & Rose the leaves are mostly 10–30 cm long. Peduncles are much shorter than leaves. Occasional. Similar places; and freshwater marshes, swamps, pools, and ponds; LA–sVA. Apr.–July.

Scotch Lovage
Ligusticum scothicum L.

[153]
Stout, erect, glabrous perennial from a large aromatic root. Leaves compound, the rachis wingless. Leaflets broad and distinct, main lateral veins traceable directly to or near a tooth. Sheaths of upper leaves less than 1 cm wide. Linear bracts present below

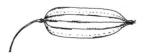

Ligusticum scothicum

the inflorescence. Fruit glabrous with the swollen style base persisting at tip. Occasional. Salt or brackish marshes, sandy or rocky shores; NY–Greenl. June–Sept.

Cow-parsnip
Heracleum maximum Bart.

[154]
Coarse, erect, rank-smelling perennial to 2.8 m tall. Stem hairy. Leaves compound. Principal leaflets 15–60 cm wide, about as long, and irregularly toothed. Petiole sheaths conspicuously expanded, 2–5 cm wide when flattened. Ovary and fruit usually hairy, not bristly or spiny. Occasional; freshwater marshes, depressions, protected beaches; NY–Nfld. May–Aug. Syn.: *H. lanatum* Michx.

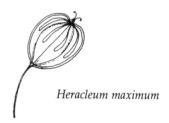

Heracleum maximum

Wild Carrot; Queen-Anne's-lace
Daucus carota L.

[155]
Erect, usually branched, bristly-hairy biennial to 2 m tall from a strong taproot. Leaves pinnately dissected, often so deeply as to appear compound, the segments narrow. Inflorescence a compound umbel, rounded when in flower but concave earlier and when in fruit, with a whorl of narrow once-pinnate leaves at base. Wings of fruit with 12 or more prickles. Common. Roadsides, fencerows, waste places, fields, grasslands, thin woods; VA–Greenl. May–Sept.

ERICACEAE
Heath Family

Indian-pipe
Monotropa uniflora L.

[156]
Plants white, yellowish, reddish, pink, or lavender, completely lacking chlorophyll, living on organic matter, probably with the aid of fungi. Flowers are solitary at end of stem and drooping. Fruits are capsules that become erect as they develop. Plants blacken rapidly as they dry. Occasional. In pine or broadleaf woods; under large shrubs; GA–Nfld. June–Oct.

PRIMULACEAE
Primrose Family

Water Pimpernel
Samolus parviflorus Raf.

[157]
Slender glabrous perennial to 50 cm tall, sometimes unbranched in small plants, as in the photograph, to freely branched in large plants. Leaves entire, mostly basal to evenly distributed. Pedicels are peculiar in having a small bract about midway. Petals white, united, the tube very short. Hypanthium fused to base of ovary. Fruit a globose capsule 2–3 mm long. Not easily confused with any other species. Common. Moist to wet places, fresh or brackish—swales, sloughs, pond and marsh margins, ditches; TX–NS. Apr.–Oct.

Loosestrife
Lysimachia quadrifolia L.

[158]
Rhizomatous perennial to 70 cm tall. Stems 4-angled. Leaves in whorls of 4–5, uncommonly less or more. Flowers axillary. Corolla united, yellow, streaked with black and reddish purple at base. Fruit a subglobose

capsule 2.5–3.5 mm long and with few seeds. Occasional. Dry to moist places—thin woods or shrub, swales, shores of ponds and freshwater marshes; VA–NH. May–Aug.

Another species with similar axillary flowers that may be encountered is *L. ciliata* L. It has opposite-petioled leaves. Petioles ciliate hairy. The blades are ovate to lanceolate-ovate, 5–15 cm long. Rare. Moist to wet places—woods, shrub areas, freshwater marshes; nVA–Nfld. June–Aug.

Swamp Loosestrife
Lysimachia terrestris (L.) B. S. P.

[159]
Rhizomatous perennial to 1 m tall. Leaves opposite, those on about the lower ⅓ of stem scalelike, the others lanceolate to narrowly elliptic, 3–10 cm long, usually over 8 mm wide. Flowers in terminal racemes with conspicuous narrow bracts at base of each whorl of flowers. Petals united at base. Occasional. Moist to wet places—swamps, thin woods, freshwater marshes, pond margins; nVA–Nfld. May–Aug.

Starflower
Trientalis borealis Raf.

[160]
Perennial to 25 cm tall, spreading by long slender stolons. Leaves in a single whorl, from the axils of which appear 1 or more flowers on long pedicels. Corolla spreading, flat, with a very short tube and usually 7 lobes. Base of filaments joined in a membranous ring. Occasional. Broadleaf and pine woods, peaty soils, shrub thickets, bogs; NY–Nfld. May–July.

Sea-milkwort
Glaux maritima L.

[161]
Succulent, erect to spreading, usually much-branched perennial, sometimes forming dense masses over 1 m across. Leaves opposite, entire, crowded. Flowers solitary in leaf axils, sessile or nearly so, lacking a co-

rolla. Sepals petal-like, united at base, white to pink or red. Occasional. Beaches, brackish or salt marshes; nVA–Nfld. June–July.

Scarlet Pimpernel
Anagallis arvensis L.

[162]
Low spreading, much-branched, rarely erect annual with ovate opposite sessile leaves. Flowers solitary in leaf axils, on long pedicels. Corolla as long as calyx or longer, usually scarlet but sometimes white or blue. Fruit a many-seeded globose capsule 3–6 mm long with a lid opening near the middle at an even horizontal circular line. Occasional. Waste places, roadsides, around buildings and parking lots; TX–AL; MD–Greenl. Apr.–frost.

PLUMBAGINACEAE
Leadwort Family

Sea-lavender; Marsh-rosemary
Limonium carolinianum (Walt.) Britt.

[163, 163a]
Sea-lavender is easy to recognize because it is confined to upper parts of salt or brackish marshes, where it is the only plant whose leaves are large, 4–17 cm long, and usually entirely in basal rosettes. Inflorescences are much-branched panicles, often large, to 50 cm across. White sepals and lavender to purple corollas are an unusual combination aiding in recognition. Common. Salt to brackish marshes; TX–NF. July–Oct.

Separated by some into two species: *L. carolinianum* (TX–NY) with obtuse calyx lobes; *L. nashii* Small (TX–Nfld) with acute to acuminate calyx lobes.

LOGANIACEAE
Logania Family

Polypremum
Polypremum procumbens L.

[164]
Glabrous perennial with several to many radiating prostrate to ascending branches, often forming a circular mass, to 70 cm across in vigorous plants. Leaves entire, awl-shaped to linear, opposite, connected at base by a fine line. Flowers solitary and sessile in leaf axils and at ends of branches. Corolla regular. Petals 4, united at lower ⅔. Stamens 4, very short. Ovary superior. Carpels 2. Fruit a capsule, notched at apex, many-seeded, 1.5–2.5 mm long. Common. Usually in well-drained places—stable dune areas, meadows, thin woods, roadsides, waste places; eTX–NY. May–Oct.

GENTIANACEAE
Gentian Family

Common Marsh-pink
Sabatia stellaris Pursh

[165]
Sabatia species are unusual in having 1-celled ovaries with 2 parietal placentas. Carpels 2.

Annual to 60 cm tall with opposite, sessile, entire leaves and no stipules. Blades elliptic to elliptic-lanceolate or linear, 15–40 mm long. Branches almost always alternate. Calyx united at base, lobes linear to almost filiform, usually about ⅔ as long as corolla. Corolla united, pink except for the eye, rarely white, lobes 5 and 10–15 mm long. Ovary superior. Common. Pond margins, salt and brackish marshes, swales, edge of sloughs; LA–sMA. July–Oct.

S. campanulata (L.) Torr. has a similar aspect but is a perennial from a short much-branched underground stem, has linear to narrowly lanceo-

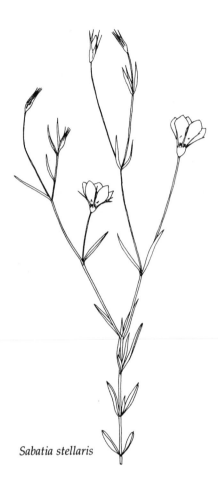

Sabatia stellaris

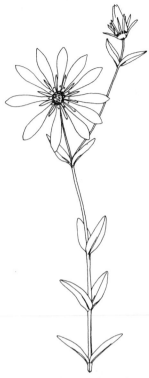

Sabatia dodecandra

late leaves, and sepals about as long as corolla. Occasional. Pond margins, swales, edge of sloughs, bogs, wet grasslands; LA–MA. May–Aug.

Large Marsh-pink
Sabatia dodecandra (L.) B. S. P.

[166]
Perennial to 1 m tall, with slender to coarse rhizomes. Branches usually alternate and restricted to upper third of stem. Leaves opposite, wider than diameter of stem, blades linear to elliptic or lanceolate. Flowers long-pedicelled (3–11 cm). Calyx lobes not hyaline-margined. Corolla united at base, the lobes 7–13, pink or rarely white. Occasional. Salt, brackish, or freshwater marshes; ditches, pond margins; TX–CT. June–Sept.

S. kennedyana Fern. has similar flowers but pedicels often shorter (0.5–6 cm) and calyx lobes hyaline-margined. Rare. Pond margins, freshwater marshes; nSC–sNC; RI–MA; sNS. June–Sept.

MENYANTHACEAE
Bogbean Family

Floating-heart
Nymphoides aquatica (Walt.) O. Ktze.

[167]
Perennial from a thick rhizome. Leaf blades floating, resembling those of some water lilies, nearly circular, 5–20 cm across, veins prominent, undersurface pebbly and purple. What seems to be a long petiole is mostly stem, on the terminal end of which develops an umbel of flowers and 1 leaf with a short petiole. After flower-

ing, tubers to 2 cm long and 4 mm thick usually develop and hang downward under the umbel. Occasional. In water of ponds, lakes, sloughs, and slow streams; eTX–sNJ. Apr.–Sept.

The similar *N. cordatum* (Ell.) Fern. has ovate, usually smaller (3–7 cm long) leaf blades that are not pebbly beneath. The tubers are much elongated and very slender. Rare. Similar places; CT–Nfld. Apr.–Sept.

APOCYNACEAE
Dogbane Family

Indian-hemp
Apocynum cannabinum L.

[168]
Perennial to 1.2 m tall, juice milky. Leaves spreading or ascending. Corolla 2–5 mm long, cylindrical, with erect or only slightly divergent lobes. The fruit from any one flower consists of a pair of slender follicles 12–22 cm long. Each seed with a tuft of long silk hairs at apex. Members of genus are reported to be poisonous when eaten. Common. Thin woods or in open—sandy flats between stable dunes, thin scrub, waste places, roadsides; GA–NH. May–Aug.

A. androsaemifolium L., which is similar, has spreading to drooping leaves, a bell-shaped corolla 5–10 mm long with the lobes spreading or curved backward. Common. Roadsides, thin woods, thin scrub; NY–NS. June–Aug.

ASCLEPIADACEAE
Milkweed Family

Red Milkweed
Asclepias lanceolata Walt.

[169]
Milkweeds are perennial and the juice is milky except in two species. The flowers are in umbels and have a complicated structure, although some characteristics are easily understood and helpful in identification to species. Flowers consist of an outermost calyx with 5 small reflexed lobes, usually hidden by the 5 reflexed lobes of the corolla, which is red in the accompanying photograph. The orange structures above the red corolla are hoods that together form the crown. The crown may be elevated on a column as seen in the photograph. Each hood contains a cavity from which a needlelike horn may protrude. Fruits are follicles; in most species they are erect and on deflexed pedicels. Seeds, except in 1 species, bear a tuft of silky hairs. Several species are known to be poisonous when eaten raw; it is likely that most, if not all, species are toxic.

This species may grow to 1.2 m tall. The stem is erect and rarely branched. The leaves are linear to narrowly lanceolate and in 3–6 pairs. There are 1–6 umbels, corolla red, crown usually orange, and horns hidden. The fruits are erect on deflexed pedicels. Occasional. Ponds, sloughs, wet savannas and pine barrens, freshwater or brackish marshes; eTX–sNJ. May–Sept.

A. rubra L. has a similar aspect and the hoods are orange but the corolla is dull purplish to lavender and leaves are lanceolate to ovate-lanceolate. Rare. Freshwater marshes, bogs, swamps, wet meadows; eTX–wFL. June–Aug.

Butterfly-weed; Chigger-weed
Asclepias tuberosa L.

[170]
Plants roughish-hairy. Stems 1 to several from a thick root; erect, ascending, or decumbent; usually branching at the top into 2–5 parts, these sometimes elongated and spreading and each bearing 1–6 umbels. The sap is not milky. Leaves abundant, alternate, linear to elliptic, obovate, oblanceolate, or hastate. Corolla and crown usually orange but varying to yellow or red. Either the crown or the corolla may be darker than the other. Com-

mon. Dry places, thin woods or in the open; nGA–NH. May–Aug.

Sandhill Milkweed
Asclepias humistrata Walt.

[171]
Stems stiff and spreading, 1 to several from a deep narrowly fusiform root. The spreading habit, 5–10 close pairs of broad sessile clasping leaves, and abundant milky juice make this species distinctive. Other prominent features include the tan-colored flower buds, a nearly white crown, and erect fruits on deflexed pedicels. Plants often grow in very hot and dry places without wilting. Common. Dry sandy soils—open pinelands, scrub areas, savannas; eLA–sNC. Mar.–July, occasionally to Sept.

Common Milkweed
Asclepias syriaca L.

[172]
Plants to 2 m tall. Leaves opposite, widely elliptic to ovate-elliptic, finely hairy below. Corolla 8–10 mm long, greenish to nearly purple. Hoods 6–8 mm long, pale purple. Horns shorter than hood. Fruit erect on drooping pedicels, surface with many slim to conic structures 1–3 mm long. Common. Open, usually dry places—fields, roadsides, waste places, stable dune areas; nVA–NS. June–Aug.

Swamp Milkweed
Asclepias incarnata L.

[173]
Plants to 1.5 m tall. Leaves numerous, petioled, linear-lanceolate to ovate-elliptic, hairy beneath, 1–3 cm wide. Hoods 3–4 mm long. Corolla and crown pink to rose-purple. Horns conspicuously protruding beyond the hoods. Fruits erect on erect pedicels. Common. Moist to wet places—marshes, sloughs, swamps, ditches; nNC–NS. July–Sept.

Sand-vine
Cynanchium angustifolium Pers.

[174]
Perennial twining vine with opposite, sessile, narrowly linear leaves. Flowers in axillary peduncled umbel-like clusters. Flowers 4–5 mm long. Calyx lobes 1–2 mm long. Fruit a follicle 4–7 cm long, 5–7 mm thick. Seeds bearing tufts of long silky hairs at their tops. Common. Salt and brackish marshes, salt flats, shell middens, grasslands, low open scrub, hammocks; TX–NC. May–Sept. Syn.: *C. palustre* (Pursh) Heller.

C. scoparium Nutt. has the same general aspect but leaves with a short slim petiole, flower clusters small and with very short peduncles, flowers about 2 mm long, and calyx lobes about 1 mm long. Rare. Scrub areas, hammocks; eFL–sSC. May–Oct.

Spiny-pod
Matelea carolinensis (Jacq.) Woods.

[175]
Seaside *Matelea* species are perennial twining vines with opposite petioled cordate to sagittate leaves. Flowers are in umbel-like clusters arising from leaf axils. Fruits are pointed follicles with many seeds, which at their tips bear a tuft of long silky hairs.

As the common name of this species suggests, fruits of this species are spiny, although not especially sharp. Petals are 10–15 mm long and spreading. Occasional. Usually in well-drained soils of woods, thickets, scrub; GA–SC. Apr.–June.

M. gonocarpa (Walt.) Shinners (Angle-pod) is the only other *Matelea* likely to be encountered. It has a smooth 5-angled fruit. Corolla yellow to orange-brown, maroon, or nearly black. Occasional. Usually in moist places in woods or thickets; FL–sVA. May–July. Syn.: *M. suberosa* (L.) Shinners.

CONVOLVULACEAE
Morning-glory Family

Dodder; Love-vine
Cuscuta gronovii Willd.

[176]
Dodders are annual, twining, parasitic, yellowish to orange vines without chlorophyll, with roots absent on mature plants, attached to host plants by rootlike suckers. Leaves represented by minute scales. Flowers small, corolla united below and lobed above. Fruit a globose to ovoid capsule. Identification to species is often difficult. We have recorded 5 species in seaside habitats. We include the one most often encountered.

In this species the calyx is 5-lobed, not 5-angled, shorter than corolla tube, lobes rounded at tip. Corolla with 5 lobes. Styles separate. Capsule ovoid, breaking open along irregular lines. Common. Parasitic on a variety of species, usually growing in moist to wet habitats—freshwater and brackish marshes, sloughs, swamps, pond margins; TX–NS. July–Oct.

Pony-foot
Dichondra carolinensis Michx.

[177]
Perennial with prostrate spreading stems that root at nodes. Often forms dense masses and is sometimes used as a ground cover. Flowers axillary, solitary. Fruit a 2-lobed, 2-seeded capsule. Common. Moist to wet places, tolerating short-time flooding—broadleaf woods, pinelands, roadsides, pond margins, swales, lawns; eTX–VA. Mar.–Sept.

Hedge Bindweed
Calystegia sepium (L.) R. Br.

[178]
Calystegia species are perennial twining or occasionally trailing or erect vines, the calyx concealed by 2 large bracts, and the fruits 4-seeded capsules.

This species trails or twines, the flowers axillary and along most of stem, stems and leaves glabrous to lightly and finely hairy. Bracts 15–35 mm long. Corolla pink with white stripes to entirely white. Stamens 20–35 mm long. A variable species with several varieties based on leaf and bract shapes and flower color. In many areas a noxious weed. Common. Borders of salt marshes, brackish areas, splash zone of beaches, fields, roadsides, thin woods; eTX–Nfld. May–Oct. Syn.: *Convolvulus sepium* L.; *C. repens* L.; *C. americanus* (Sims) Greene.

Railroad-vine
Ipomoea brasiliensis (L.) Sweet

[179]
Seaside *Ipomoea* are twining or trailing vines with the calyx not hidden by bracts. There is one stigma, which is 2-lobed in some species. Mature fruits are dry, 2- to 4-celled and 2- to 6-seeded.

Railroad-vine is a trailing fleshy glabrous perennial, rooting at the nodes. We have seen stems as long as 31 m long trailing across dunes. Leaves are unlobed, up to 11 cm wide, corolla funnel-shaped. Common. Drift area of beach, dunes, overwash flats; TX–sSC. June–Nov. *I. pes-caprae* (L.) R. Br. as treated in other manuals.

Fiddle-leaf Morning-glory
Ipomoea stolonifera (Cyr.) Gmel.

[180]
Trailing fleshy glabrous perennial with stems rooting at nodes. Most leaf blades lobed near base, sometimes deeply so. Flower stalks about as long as leaves. Common. Drift areas of beach, dunes, overwash flats; TX–sNC. Apr.–Nov.

Arrow-leaf Morning-glory
Ipomoea sagittata Poir.

[181]
Glabrous twining or trailing perennial with arrow-shaped leaves. Flowers solitary from leaf axils. Corolla funnel-shaped, 55–75 mm long. Anthers 5–7 mm long. Common. Freshwater

and brackish marshes, swales, sloughs, edge of ponds, stable dune areas; TX–NC. Apr.–Oct.

Coastal Morning-glory
Ipomoea trichocarpa Ell.

[182]
Twining perennial from a branched root. Leaves usually lobed, blades 4–10 cm long. Inflorescence of 1–8 flowers. Pedicels glabrous. Sepals unequal, 7–11 mm long, ciliate on margin and hairy at base. Corolla funnel-shaped, 28–55 mm long, pink to purple or rarely white. Anthers 1.5–3.2 mm long. Stigma 2-lobed. Mature fruit 2-celled and 4-seeded. Common. Stable dune and scrub areas, pond and slough margins, waste places; eTX–NC. May–Oct. Syn.: *I. carolina* (L.) Pursh.

Wild Potato-vine
Ipomoea pandurata (L.) Mey.

[183]
Perennial from a deep vertical tuberous root that may weigh up to 30 lbs. (14 kg). Plant glabrous or with a few small hairs. Leaf blades 2–10 cm long, cordate-ovate to somewhat fiddle-shaped. Corolla 6–8 cm long, always with a reddish purple center. Anthers 5–7 mm long. Occasional. Thin woods, thickets, waste places, roadsides; eTX–CT. May–Sept.

POLEMONIACEAE
Phlox Family

Annual Phlox
Phlox drummondii Hook.

[184]
Growing from seed each year, to 70 cm tall. Upper stem leaves often alternate, the lower opposite. Glandular hairs on stem, leaves, calyx, and corolla tube. There is a great variety of cultivated forms differing especially in color and shape of the corolla. Escapes are frequent. Native in TX. Common. Sandy soils—stable dune

areas, roadsides, meadows, grasslands, thin woods; TX–sMD. Mar.–July.

BORAGINACEAE
Borage Family

Marsh Heliotrope
Heliotropium curassavicum L.

[185]
Glabrous prostrate to ascending annual. Leaves alternate, entire, mostly sessile. Flowers in terminal curved spikelike groups, these usually in pairs, becoming nearly straight as fruit matures. Corolla white with a yellow eye, regular. Stigma sessile. Ovary shallowly 4-lobed. Mature fruit 4-lobed, 2–2.5 mm high, breaking into 4 one-seeded sections. Common. Seashores, brackish to salt marshes, overwash flats, brackish sloughs and swales; TX–DE. May–Oct.

VERBENACEAE
Vervain Family

South American Vervain
Verbena bonariensis L.

[186]
All *Verbena* species have opposite simple leaves. Flowers are in terminal spikes and subtended by bracts. Corolla with a slender tube expanding abruptly into slightly irregular lobes. Stamens attached to the upper part of the corolla at 2 levels. Fruit an aggregate of 4 nutlets held closely together by the tube of the united sepals. Most *Verbena* flowers are small. Identification of most species is not easy. We have recorded 9 species in seaside habitats but most are unlikely to be encountered.

This species is a perennial to 2.5 m tall with square stems. Leaves sessile; blades lanceolate to elliptic-lanceolate, the bases little narrowed and partly clasping stem. Flowers are in dense spikes 5–6 mm broad when fully developed. Common. Roadsides, waste places, marsh edges, swales, ditches, clearings; eTX–NC. Apr.–Oct.

V. brasiliensis Vell. is quite similar but leaves are acutely narrowed at base and the spikes are 4–5 mm broad when fully developed. Occasional. Similar places; eTX–sVA. Mar.–Nov.

Verbena bonariensis

Verbena brasiliensis

Verbena halei

Verbena scabra

Texas Vervain
Verbena halei Small

[187]
Some seaside vervains have flowers in slender spikes that gradually elongate as the fruits develop, until fruits are well spaced or scarcely overlapping. Texas vervain is one of these. It is a perennial to 80 cm tall with square stems and opposite petioled pinnately dissected leaves. Calyx 2.5–3.5 mm long, the subtending bract about half as long. Corolla 4–6 mm broad. Occasional. Open sandy areas—between stable dunes, meadows, fields, roadsides; TX–AL. Mar.–frost.

V. scabra Vahl also has slender

spikes with well-spaced fruits. Leaves petioled, 3–13 cm long, narrowly ovate, serrate-dentate. Calyx 1.5–2 mm long, tips of the sepals curving over nutlets, producing a beak. Corolla about 2 mm broad. Occasional. Margins of freshwater or brackish marshes, wet woodlands, swales, shell mounds; TX–sVA. Apr.–Oct.

Frog-fruits
Phyla nodiflora (L.) Greene

[188]
Perennial, usually with rooting prostrate or decumbent stems. Leaf blades oblanceolate to spatulate or obovate with apex obtuse to rounded, rarely acute, with 1–7 teeth above the middle on each side. Flowers many in a light head that is longer than broad as the seeds become mature. Common. Sandy, usually wet soils— swales, sloughs, pond margins, ditches, low pinelands; TX–sVA. Apr.–frost. Syn.: *Lippia nodiflora* (L.) Michx.

P. lanceolata (Michx.) Greene is similar but has taller ascending to erect stems as well as the prostrate ones. Leaf blades mostly lanceolate to lanceolate-elliptic, with a strong taper to an acute apex, teeth 5–11 on each side extending below the middle. Occasional. Similar places; eTX–AL; eFL–sNJ. May–frost. Syn.: *Lippia lanceolata* Michx.

LAMIACEAE
Mint Family

Germander; Wood-sage
Teucrium canadense L.

[189]
Members of the mint family have opposite simple leaves, usually square stems, corolla irregular and 2-lipped, petals united, ovary superior, style in most genera arising from the central depression of the 4 lobes of the ovary, fruit usually 4 nutlets nestled in the persistent calyx tube.

Germander is an erect rhizomatous perennial to 1.5 m tall. Corolla purplish, pink, or cream-colored. Upper 4 lobes of corolla nearly equal, oblong, turned forward so that there seems to be no upper lip. Lower lip very prominent. Anther-bearing stamens 4. Ovary 4-lobed, not deeply 4-parted as in most mints. Fruit of 4 nutlets fastened at their sides. Often confused with *Stachys* species, which have a distinctly 2-lipped corolla. Common. Moist to wet places in thin woods or open—freshwater marshes, edge of brackish marshes and ponds, meadows, sloughs, ditches, wet grasslands; TX–NS. May–Sept.

Blue-curls
Trichostema dichotomum L.

[190]
Annual to 80 cm tall, bushy-branched. Leaves elliptic to elliptic-lanceolate, 2–7 cm long, blades less than 5 times as long as wide, the lower ones often falling under dry conditions. Flowers abundant. Calyx irregular. Lower lip of corolla drooping. The 4 stamens are bluish and strongly arched, thus the common name. Common, rare along Gulf. Well-drained soils—between stable dunes, roadsides, thin woods and pinelands, fields; eTX–sME. July–frost.

Northern Skullcap
Scutellaria epilobiifolia A Hamilt.

[191]
Weak-stemmed perennial to 80 cm tall from slender creeping rhizomes. Stems with fine retrorse hairs. Leaves sessile or with petioles to 4 mm long, blades 2–4 times as long as wide. Flowers in axils of foliage leaves, mostly solitary. Calyx 2-lipped, closed in fruit, with a helmetlike projection on the upper side. Corolla 15–25 mm long. Occasional. Moist to wet places—meadows, swales, marsh shores, thickets, swamps; DE–NS. June–frost.

Common Skullcap
Scutellaria integrifolia L.

[192]
Erect perennial to 80 cm tall, simple or branched above; finely hairy. Lower leaves with slender petioles, blades ovate and 15–40 cm long. Upper leaf blades narrowly lanceolate to oblanceolate, entire or nearly so, 2–6 cm long. Flowers in terminal racemes. Calyx with a conspicuous protuberance on upper side. Corolla 18–28 mm long. Common. Usually moist places—thin woods, pine barrens, meadows, margins of sloughs and ponds, roadsides, ditches; eTX–NY. Apr.–July.

Heal-all
Prunella vulgaris L.

[193]
Perennial to 80 cm tall with short branches anywhere below the central flower cluster. Flowers in dense, nearly globose spikes, becoming cylindrical as fruits mature, the spikes 1 cm wide excluding corollas. Corolla white to pink and purple. Pollen-bearing stamens 4. Common. Roadsides, lawns, flower bed borders, paths; GA–Nfld. Apr.–frost.

Hemp-nettle
Galeopsis tetrahit L.

[194]
Annual to 1.2 m tall. Stems with spreading hairs. Flowers in 2–6 dense whorls in each inflorescence. Calyx lobes equal or nearly so, erect to spreading, spine-tipped. Corolla 15–25 mm long, upper lip hoodlike, lower lip with 2 yellow spots. Stamens 4, longer than corolla tube. Common. Roadsides, waste places, around buildings, disturbed areas; NY–Nfld. June–Sept.

Henbit
Lamium amplexicaule L.

[195]
Annual or winter-annual to 35 cm tall, branching from the base, most branches arched and ascending.

Leaves 1.5 times or less as long as wide. Flowers in axillary clusters. Corolla 12–18 mm long, upper lip hoodlike. Common. Roadsides, waste places, disturbed soils, yards and borders; parking areas; TX–Greenl. Mild winter–June.

Betony
Stachys floridana Shuttlew.

[196]
Erect perennial to 80 cm tall from slender rhizomes that are tuberous-thickened at intervals. Leaves with petioles to 45 mm long, blades lanceolate to lanceolate-ovate. Flowers in axils of bracts in long terminal groups. Corolla 10–11 mm long, upper lobe hoodlike. Occasional. Usually in wet places—lawns and bor-

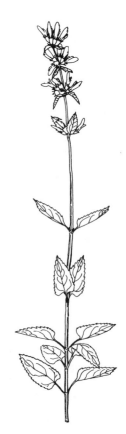

Stachys floridana

ders, roadsides, around buildings, thickets; eTX–sVA. Mar.–July.

Lyre-leaved Sage
Salvia lyrata L.

[197]
Perennial to 80 cm tall. Leaves basal or mostly so. Usually only 1 stem but sometimes 2 to several from the base and occasionally 2 leafless branches from the upper part. Pollen-bearing stamens 2. Common. In thin woods or open—yards and borders, roadsides, between stable dunes, along paths; eTX–CT. Frost-free periods.

Scarlet Sage
Salvia coccinea Juss. ex J. Murr.

[198]
Perennial to 80 cm tall. Leaves with petioles 5–40 mm long; blades ovate, cordate at base. Calyx 2-lipped. Corolla 25–30 mm long. Anther-bearing stamens 2, longer than corolla tube. Occasional. Thin woods, roadsides, shell mounds, around buildings, waste areas; eTX–sSC. Feb.–Nov.

Salvia coccinea

Horse Mint
Monarda punctata L.

[199]
Perennial to 1 m tall. Leaves widest below the middle. Flowers in 2 or more tight clusters at end of each flowering stem, each cluster with several wholly or partially pink to lavender leaflike bracts beneath. Calyx lobes acute to acuminate, to 2 mm

long. Corolla cream to yellow, spotted with purple. Common. Dry places in thin woods or open—between stable dunes, grasslands, roadsides; eTX–NY. July–Sept.

Water-horehound
Lycopus americanus Muhl. ex Bart.

[200]
Lycopus species are stoloniferous perennials. Flowers are in compact axillary clusters. Calyx regular. Corolla 5-lobed, slightly irregular.

In this species plants are to 1 m tall, stems acutely 4-angled, with underground stolons but no aboveground runners, and the calyx lobes glandular-ciliate and subulate-tipped. Common. Moist to wet places—woods, freshwater marshes, sloughs, pond edges, ditches; VA–Nfld. June–frost.

Two similar species may be encountered: *L. virginicus* L. has rounded stem angles and obtuse to acute calyx lobes. Common. Similar places; eTX–wFL; SC–MA. July–frost. *L. rubellus* Moench. has rounded stem angles, tapered leaf bases and long-acuminate calyx lobes. Occasional. Similar places; eTX–MA. June–frost.

Bitter Mint
Hyptis alata (Raf.) Shinners

[201]
Perennial to 2.5 m tall. Stem single or with a few branches in the upper portion. Flowers in dense heads on stalks from upper leaf axils. Calyx lobes nearly equal. Corolla almost white, spotted with purple, the lower lip with a saclike lobe. Stamens 4, protruding from the corolla tube. Occasional. Moist to wet places—swales, prairies, open pinelands, pond margins, ditches; eTX–sNC. June–Oct.

SOLANACEAE
Nightshade Family

Sand Ground-cherry
Physalis viscosa L. ssp. *maritima* (M. A. Curtis) Waterfall

[202]
Physalis species are annuals or perennials. Fruits are peculiar little tomato-like berries that are enclosed in a considerably larger paper sac with only a small opening at the tip. The sepals, united and enlarged, form the sac.

Sand Ground-cherry is a perennial to 60 cm tall with small stellate hairs on stem, leaves, and part of the flower. Anthers are yellow. Common. Sandy soils—dunes, interdunes, thin scrub; LA–VA. Apr.–Sept.

P. angustifolia Nutt., a glabrous or nearly so perennial with linear leaves, is occasional in similar habitats from LA–wFL. Mar.–Aug.

Two annuals may be encountered. One is the glabrous or nearly so *P. angulata* L. with a 10-ribbed fruiting calyx. Occasional. Thin woods, roadsides, disturbed places; eTX–sVA. June–Oct. The other, *P. pubescens* L., is usually hairy and has a 5-angled fruiting calyx. Occasional. Similar places; eTX–sVA. June–Oct.

Horse-nettle; Bull-nettle
Solanum caroliniense L.

[203]
Erect simple to branching perennial to 80 cm tall from deep vigorous horizontal rhizomes. Prominent straw-colored prickles are on stems and underside of leaves. Stem and leaves also loosely covered with small 4- to 8-rayed hairs. Leaf blades usually ovate, each side with 2–5 large teeth or shallow lobes. Fruits are tomato-like berries, green at first and later yellow. They are poisonous when eaten. Common. Roadsides, around buildings and parking areas, waste places; eTX–NY. Apr.–Oct.

Spiny Nightshade
Solanum sisymbriifolium Lam.

[204]
Much-branched coarse annual to 1 m tall; the stem leaves, flower stalks, and calyx armed with flat yellow to orange spines. Leaves pinnately parted. Corolla white to violet. Mature fruits are red berries 12–16 mm across. Occasional. Roadsides, fence-rows, dikes, waste places; eTX–GA; rare north to MA. May–Sept.

Bittersweet
Solanum dulcamara L.

[205]
Unarmed glabrous to short-hairy perennial climbing or scrambling to 3 m tall. Often not dying to ground; then a woody plant. Leaves petioled, some simple and unlobed, others with a pair of smaller basal lobes or leaflets. Corolla light blue to violet, the lobes 5–9 mm long, becoming reflexed. Fruits are red berries 8–11 mm long, poisonous. Occasional. Thickets, thin woods, fencerows, around buildings; NJ–NS. May–Sept.

Black Nightshade
Solanum pseudogracile Heiser

[206]
There probably are 4 species of black nightshade in seaside habitats and they are often difficult to identify. Most manuals do not agree as to the species involved or their characteristics. As a group they can be easily recognized. All are annuals; stems and leaves lack prickles; leaf blades are entire to sinuate or coarsely dentate, not deeply cut; flowers are in umbel-like clusters or in short racemes on peduncles that peculiarly appear to arise from the internodes (fundamentally they do not). Corollas are white, 6–10 mm wide. Fruits are berries, like tiny tomatoes, and 5–13 mm in diameter.

This species is finely hairy, anthers over 2 mm long, and fruit lacking hard granules. Common. Dune and drier interdune areas, maritime woods, edge of brackish marshes; eMS–sNC. Apr.–frost. Syn.: *S. gracile* Link.

S. americanum Mill. is glabrous or nearly so, anthers less than 2 mm long, and fruit contains hard granules, usually 4 or more. Common. Woods, margins of ponds, disturbed soil, borders of yards, waste places; TX–MS; nNC–sME. Mar.–frost.

SCROPHULARIACEAE
Figwort Family

Woolly Mullein; Flannel-plant
Verbascum thapsus L.

[207]
Seaside members of this family are herbs with usually circular stems, simple leaves, weakly to strongly irregular corollas with 2 lips, usually 4 stamens, and a superior 2-carpelled ovary with a terminal style. Fruits are capsules.

This species is a densely woolly biennial to over 2 m tall. Stem is upright, stout, and mostly unbranched. Leaves have feel of thick flannel, in a basal rosette only during the first year; those of second year to 40 cm long, bases of upper leaves extending down sides of stem. Corollas yellow, or rarely white. Terminal inflorescence to nearly 1 m long, that in photograph shortened by adverse conditions of the overwash area where the plant was growing. Common. To be expected in almost any place except dense woods, wet places, and active dunes; TX–NS. Mar.–Nov.

Toadflax
Linaria canadensis (L.) Dum.-Cours.

[208]
Slender glabrous biennial or winter annual to 75 cm tall, with a rosette of prostrate stems to 10 cm long. Upper leaves alternate, linear; those on the prostrate stems opposite or whorled or nearly so, and wider. Flowers in 1 to several racemes. Corolla blue to purple or rarely white, with a slender

Linaria canadensis

spur at base, the spurs 5–9 mm long. Common. Stable dune areas, fields, yard gardens, roadsides; eTX–sNS. Mar.–Sept.

L. floridana Chapm. is similar but the spurs are under 1 mm long and plants only 40 cm tall. Occasional. Dry sandy soils in thin woods or open; MS–GA. Mar.–Apr.

Butter-and-eggs
Linaria vulgaris Mill.

[209]
Perennial with 1 to several glabrous ascending stems to 1.3 m tall. Leaves very numerous, linear, narrowed to a petiolelike base. Flowers in terminal racemes. Corolla yellow with an orange throat and a conspicuous spur. Occasional. Gravel and sandy shores, waste places, roadsides, around buildings; nVA–Greenl. May–Sept.

Smooth Gratiola
Gratiola virginiana L.

[210]
Erect, succulent, glabrous or nearly so
annual to 40 cm tall. Stem simple or
with a few branches. Leaves opposite;
blades narrowed at base, not clasping. Flowers on axillary pedicels to 6
mm long and 0.6–0.9 mm in diameter. Calyx lobes 5, longer than tube.
Corolla nearly white, 9–14 mm long.
Pollen-bearing stamens 2, sterile stamens very small or missing. Occasional. Swales, ditches, margins of
ponds and sloughs, freshwater
marshes; eTX–NY. Mar.–Oct.
   The similar *G. neglecta* Torr. has filiform pedicels 10–25 mm long and
0.3–0.4 thick at midpoint. Corolla 8–
12 mm long. Rare. Similar places;
VA–NY. Apr.–June.

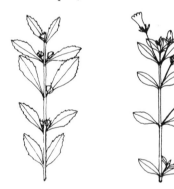

Gratiola virginiana   *Gratiola neglecta*

Hairy Gratiola
Gratiola pilosa Michx.

[211]
Stiff, erect, usually unbranched hairy
perennial to 75 cm tall. Stem 1–2 mm
in diameter. Leaves opposite, blades
ovate to ovate-lanceolate, entire or irregularly serrate. Flowers and fruits
sessile or nearly so. Calyx lobes 5.
Corolla equal to or slightly longer
than calyx. Pollen-bearing stamens 2,
sterile stamens absent or minute. Occasional. Dry to wet places—thin
woods, ditches, freshwater marshes,
pond and slough edges; eTX–DE.
June–Oct.

Gratiola pilosa

Blue Water-hyssop
Bacopa caroliniana (Walt.) Robins.

[212]
Creeping or floating perennial to 30
cm tall, often forming extensive mats.
Lemon-scented when crushed.
Leaves opposite, ovate to broadly elliptic, apex obtuse to rounded, with
3–7 palmate veins. Flowers solitary in
leaf axils. Calyx lobes 5, longer than
tube. Corolla 9–11 mm long, the lobes
slightly different in size and shape.
Pollen-bearing stamens 4. Occasional.
Shallow water and moist edges of
sloughs, ponds, and ditches; eTX–
DE. Apr.–Oct. Syn.: *Hydrotrida caroliniana* (Walt.) Small.

Smooth Water-hyssop
Bacopa monnieri (L.) Penn.

[213]
Plant not lemon-scented. Stems glabrous, succulent, creeping, usually
branching extensively and forming
large glossy green mats. Flowering
branches decumbent or ascending to
20 cm tall. Leaves opposite; blades
spatulate to cuneate-ovate, 6–17 mm
long, with 1 vein. Pedicels, at least in
fruit, exceeding leaves. Corolla bell-
shaped, white to light purple, the
lobes slightly unequal. Pollen-bearing
stamens 4, small. Common. Swales,
sloughs, ditches, pond edges, fresh
or brackish marshes, sand flats; TX–
sVA. Apr.–frost.

Micranthemum
Micranthemum umbrosum (Walt.) Blake

[214]
Low, creeping, matted glabrous plants forming dense bright green carpets to 1.5 m across. Sometimes attached under water to roots, stems, or dead branches of woody plants. Leaves opposite; blades orbicular, 4–9 mm broad and long or, when submersed, elongated to 15 mm. Flowers minute, under 2 mm long, solitary in leaf axils, usually only 1 per node. Calyx lobes 4. Corolla white, 4-lobed, about 1.5 mm long. Common. Wet places—edges of and in streams, ponds, ditches, sloughs, freshwater or brackish marshes; eTX–sNC. Apr.–Oct.

Corn Speedwell
Veronica arvensis L.

[215]
In seaside speedwells, leaves on the lower parts of the plant are opposite, those below each flower are alternate. Flowers axillary, solitary. Calyx lobes 4, quite distinct. Corolla lobes nearly equal, none concave. Stamens 2. Fruits 2-carpelled, flattened, obcordate, hairy on the angles.

V. arvensis is a hairy decumbent to erect annual to 30 cm tall. Lower leaves ovate, less than three times as long as wide, and palmately veined. Corolla light blue. Common. Roadsides, around buildings, waste places, thin woods; eTX–Greenl. Mar.–June.

V. peregrina L. (Purslane; Speedwell) has a similar aspect but is glabrous, the lower leaves are oblong to obovate and twice or more as long as wide. Corolla white. Common. Similar places; LA–NS. Mar.–Aug.

Thymeleaf Speedwell
Veronica serpyllifolia L.

[216]
Decumbent to erect perennial to 20 cm tall, often forming mats. Bracts subtending lowest flowers decidedly smaller than the widely ovate vegetative leaves. Flowers decidedly smaller than the widely ovate vegetative leaves. Flowers on pedicels 2–5 mm long. Corolla white to pale blue with deeper-colored stripes. Style as long as the fruit. Occasional. Thin woods, roadsides, around buildings, low open areas; nVA–Greenl. Apr.–Aug.

Gerardia; False-foxglove
Agalinis fasciculata (Ell.) Raf.

[217]
Gerardia species have opposite entire filiform to linear leaves, often with leaf clusters in the axils. Flowers in terminal, sometimes short, racemes. Sepals 5, lobes alike and shorter than the conspicuous tube. Corolla with a bell-shaped tube and nearly equal lobes, the lower lobe on the outside when in bud. Stamens 4, in 2 unequal pairs. We have recorded 9 species in seaside habitats. Identification of most species is difficult. The 3 most common are included here.

This species is a profusely branched annual to 1.2 m tall. Stems finely rough to touch. Often with clusters of leaves in axils of the main leaves. Pedicel shorter than the calyx tube. Corollas 20–35 mm long. Common. Thin woods, swales, roadsides, marshes and other low open areas, pond edges; eTX–sNC. Aug.–Oct.

A. purpurea (L.) Penn. is similar but stems are smooth or nearly so and clustered leaves absent. Common. Similar places; eTX–NS. Aug.–frost. The similar *A. maritima* (Raf.) Raf. is weakly succulent, leaves and calyx lobes obtuse, corolla 10–15 mm long. Occasional. Salt and brackish marshes, brackish swales; TX–NS. June–July.

False-foxglove
Aureolaria flava (L.) Farw.

[218]
Perennial to 2.5 m tall. Stem, leaves, and fruit glabrous. Leaves opposite, the upper entire to serrate, the lower usually pinnately lobed. Pedicels of flowers 4–25 mm long. Corolla yellow, lobes nearly alike, the lower lobes overlapping the others in bud. Stamens 4, anthers hairy. Occasional.

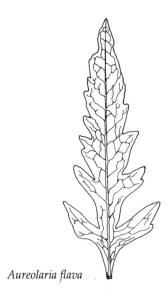

Aureolaria flava

Thin woods, pinelands; eTX–sME.
May–Sept.

A. *virginica* (L.) Penn. is generally
similar but the stems, leaves, and
fruits are hairy and lower leaves usually less lobed. Occasional. Similar
places; VA–MA. May–Aug.

Blue-hearts
Buchnera americana L.

[219]
Hairy rarely branched perennial to 80
cm tall, with a hard rootstock. Plants
turn very dark on drying. Leaves opposite, 1- to 3-veined, mostly on
lower half of stem, the lower ones
much wider than the upper, which
are sometimes alternate. Flowers in
gradually elongating spikes up to 15
cm long. Corolla purple or white.

Buchnera americana

Buchnera americana

Fruit many-seeded, ovoid to pear-
shaped. Similar to some verbenas,
but verbena fruits consist of 4 readily
separable nutlets. Common. Open
areas or thin woods, usually moist
nonsaline places; TX–sNC. Apr.–Nov.
Syn.: *B. floridana* Gand.

LENTIBULARIACEAE
Bladderwort Family

Floating Bladderwort
Utricularia inflata Walt.

[220]
Seaside plants of this genus lack roots
and probably leaves. The floating, underwater, or in-mud leaflike branches
from the main axis are most likely
stems. These are filiform and unbranched or branched. In 2 species
there is a whorl of floating structures
with swollen stalks as in the photograph of *U. inflata*. The filiform segments of the branches usually bear
small bladders that trap minute
aquatic life, which provides food.

Plants of this species float free in
the water. They have a whorl of swollen floating branches as well as several to many finely dissected underwater "leaves" that bear bladders.
The corollas are 15 mm or more broad
and the fruiting pedicels mostly recurved. Rare. Ponds, sloughs, roadside pools, drainage ditches, swamps;
eTX–NY. Feb.–Nov.

U. radiata Small is closely similar
but the corollas are under 15 mm
broad and the fruiting pedicels usually erect to spreading. Rare. Similar
places; TX–NS. Mar.–Nov.

Wiry Bladderwort
Utricularia subulata L.

[221]
Plants to 18 cm tall. Stems filiform-
wiry and bearing 1 to several minute
widely separated peltate bracts, similar ones at the base of each flower
stalk. The leaflike branches are all filiform, unbranched, and under the

Utricularia subulata

Utricularia biflora

sandy or sandy-peaty substrate. Common. Wet places—thin pine-lands, bogs, depressions, roadside ditches; eTX–NS. Mar.–Sept.

U. biflora Lam. is similar but the bracts are not peltate, being fastened at their bases; plants growing to only 10 cm tall and usually in floating tangled branches or mats. Rare. Ponds, swamps, drainage ditches, canals; eTX–MA. June–Oct.

Purple Bladderwort
Utricularia purpurea Walt.

[222]
Plants free-floating, usually in bunches or mats, abovewater stems to 10 cm tall, underwater bladder-bearing branches in whorls. Corollas deep pink to purple, 9–12 mm long. Capsule globose, 3–4 mm across. Occasional. Ponds, swamps, sloughs, ditches; MS; eFL–NS. Apr.–Oct.

PLANTAGINACEAE
Plantain Family

Common Plantain
Plantago major L.

[223]
All seaside plantains have basal leaves. Flowers in spikes on leafless stalks. Petals regular, united, papery, persisting until fruit is ripe. Fruits 2-celled with 2 to many seeds and splitting open at a horizontal circular line.

This species is a perennial with ovate leaves 2.5–10 cm wide, mostly lying on the ground. Corolla lobes less than 1 mm long. Fruit broadest at the middle and splitting there. Common. Around buildings, roadsides, fields, beside paths, swales; LA; NC–Greenl. June–frost.

P. rugelii Dcne. is similar but leaves more erect and petioles reddish, the fruit is broadest below the middle and also splits below the middle. Occasional. Similar places; VA–NS. June–frost.

Hoary Plantain
Plantago virginica L.

[224]
Winter annual to 15 cm tall, with tap-root. Leaves hairy, mostly oblanceolate, 5–40 mm wide, longest ones 2–16 cm long. Flowers in dense spikes. Common. Lawns, shrub borders, gardens, roadsides, fields, swales, waste places; eTX–sME. Mar.–June.

Buckhorn Plantain
Plantago aristata Michx.

[225]
Winter annual to 25 cm tall, with a taproot. Leaves linear or narrowly oblanceolate, to 20 cm long and 8 mm wide. Flowers in spikes to 15 cm long, conspicuous linear bract at base of each flower. Spikes more than ¼ as long as supporting stem. Common. Waste places, roadsides, dry interdune areas, yards and borders; eTX–AL; SC–ME. Apr.–Nov.

English Plantain; Ribgrass
Plantago lanceolata L.

[226]
Perennial to 60 cm tall. Leaves narrowly elliptic to lanceolate, 7–50 mm wide, to 30 cm long, with 3 to several prominent parallel veins. Spike 1–8 cm long, less than ¼ as long as supporting stem. Common. Roadsides, gardens, lawns, fields, waste places; SC–Greenl. Apr.–frost.

RUBIACEAE
Madder Family

Bluet
Hedyotis caerulea (L.) Hook.

[227]
Members of this family have opposite or whorled entire leaves with stipules between them on the stem. Stipules sometimes as large as the leaves; the stipules and leaves taken together having the appearance of whorled leaves. Flowers regular; petals united, the lobes 4 or 5, or rarely 3; ovary wholly or partly inferior.

This species is a perennial with erect stems to 20 cm tall. Stipules membranous and entire. Corolla pale blue, or rarely white, with a yellow eye, the tube 5–10 mm long. Fruit a capsule 2.0–3.5 mm broad. Common. Moist soils in open areas—meadows, swales, grassy areas, thin woods, roadsides; VA–NS. Apr.–June. Syn.: *Houstonia caerulea* L.

Trailing Bluet
Hedyotis procumbens (Walt. ex J. F. Gmel.) Fosb.

[228]
Creeping perennial or some stems decumbent. Leaves ovate to suborbicular. Flowers solitary on erect pedicels, sometimes with a few axillary. Corolla white, the tube 5–7 mm long and glabrous inside. Fruit a 2-carpeled capsule on curved pedicels. Common. Dunes, swales, thin scrub, thin woods; MS–sSC. Feb.–Apr. Syn.: *Houstonia procumbens* (Walt. ex J. F. Gmel.) Standl.

Oldenlandia
Hedyotis uniflora (L.) Lam.

[229]
Weakly erect annual to 60 cm tall, usually much shorter, varying from copiously hairy to glabrous. Stems simple to loosely branched. Stipules with prominent teeth. Flowers sessile, axillary and in terminal clusters. Corolla white, about 2 mm wide.

Common. Swales, pond margins, ditches, depressions in thin woods, on floating mats, swamps; eTX–NY. June–Oct. Syn.: *Oldenlandia uniflora* L.

Mexican-clover
Richardia brasiliensis Gomez

[230]
Hairy spreading perennial from a woody rootstock. Stipules with several threadlike projections. Flowers in dense terminal clusters. Corolla 3–4 mm long, with a short tube and 4–8 lobes. Fruits indehiscent, with stiff hairs, 3- to 4-seeded. Occasional. Roadsides, lawns, waste places, around buildings, dune areas; TX–sNC. May–frost.

R. scabra L. is similar but is an annual. The corolla is 5–6 mm long and the fruits are tuberculate. Common. Similar places; eTX–sNC. May–frost.

Rough Buttonweed
Diodia teres Walt.

[231]
Erect to spreading, usually much branched annual. Stipules with several threadlike structures. Flowers 1 per leaf axil. Calyx lobes 4, less than 4 mm long. Corolla 2–6 mm long. Fruit 2.5–4 mm long, splitting into 2 indehiscent 1-seeded segments. Common. Usually dry soils—dune areas, thin woods, roadsides, lawn borders, gardens; eTX–DE. May–frost.

Buttonweed
Diodia virginiana L.

[232]
Prostrate, ascending, or erect branching perennial. Base of leaf blades narrow to nearly cordate. Stipules membranous at base, the outer portion with 3–5 linear projections. Flowers 1, rarely 2, per leaf axil, sessile. Sepals 2, rarely 3, persistent. Corolla tube 7–9 mm long. Fruit 5–9 mm long, leathery, splitting into 2 indehiscent 1-seeded segments. Common. Usually moist places—shallow pools, pond margins, marshes, swamps, ditches, swales, sloughs; eTX–sNJ. May–frost.

Bedstraw; Catchweed
Galium aparine L.

[233]
In members of this genus leaves and stipules are alike and combined into whorls. Although stipules are involved we shall refer to all as leaves. Flowers are small. Petals 3–5, the lower portions united into a tube. Fruit when ripe separating into 2 seedlike, indehiscent, 1-seeded parts.

This species is a slender annual with weak stems to over 1 m long, often forming tangled masses. Stem angles armed with stiff retrorsely hooked bristles. Leaves usually in whorls of 8. Flowers single or in groups of 2–3. Fruits with many hairs hooked at their tips. Common. Moist or rich soils—scrub areas, woods, meadows, around buildings, waste places; TX–GA; VA–Greenl. Apr.–July.

G. triflorum Michx. has a similar aspect and similar fruits but is a perennial, the stems smooth or weakly roughened with bristles, the leaves usually in whorls of 6. Occasional. Broadleaf woods, roadsides, fields, thickets; VA–Greenl. Apr.–Sept. *G. pilosum* Ait. also has fruits with hooked bristles but the leaves are in whorls of 4 and usually under 25 mm long. Occasional. Dry woods; eTX–NH. May–Aug.

Galium aparine

Dye Bedstraw
Galium tinctorium L.

[234]
Perennial with weak reclining to ascending stems, often forming tangled masses. Stems sharply 4-angled. Leaves glabrous, in whorls of 5–6, rounded at tip. Corolla with 3, sometimes 4, obtusely tipped lobes. Mature fruits dry, black, smooth, 2–3 mm wide. Common. Moist to wet places—pond margins, ditches, sloughs, swamps, floating mats; TX–Nfld. Apr.–Sept.

G. obtusum Bigel. is similar but the stem angles are rounded, leaves in whorls of 4, and corolla with 4 acutely tipped lobes. Common. Similar habitats. TX–sNS. Mar.–July.

Purple Galium
Galium hispidulum Gray

[235]
Finely scabrous, usually much-branched perennial with stems to 60 cm long. Leaves in whorls of 4, elliptic, firm, persistently green into winter. Flowers usually in pairs, corolla white. Fruit purple, smooth, juicy when fresh. Common. Dunes, sandy scrub and pinelands, maritime woods, live oak woods; MS–sNJ. June–Aug.

CUCURBITACEAE
Gourd Family

Creeping Cucumber
Melothria pendula L.

[236]
Perennial vine, trailing or climbing by tendrils. Leaves palmately veined, varying from scarcely to strongly 3–5 lobed. Flowers unisexual, rarely bisexual. Fruit green to black, about 1 cm long, pulpy, with about 20 white seeds, poisonous when eaten. Common. Dunes, swales, scrub areas, edge of marshes, swamps; TX–NY. June–frost.

CAMPANULACEAE
Bellflower Family

Venus'-looking-glass
Triodanis perfoliata (L.) Nieuw.

[237]
Erect annual to 1 m tall, but most often half that tall or less, simple or

with a few branches at base. Stem leaves strongly clasping. Flowers sessile in leaf axils, those on the lower part of the stem not opening. Petals united, pale lavender to deep purple. Stamens 5, separate. Ovary inferior. Fruit many-seeded, opening by 3 small elongate pores at or just below the middle. Common. Yards, gardens, roadsides, waste places; eTX–sME. Apr.–July. Syn.: *Specularia perfoliata* (L.) A. DC.

T. biflora (R. & P.) Greene is similar but the stem leaves are barely clasping and the openings in the fruit are near the top. Rare. Similar places; eTX–sVA. Mar.–June. Syn.: *Specularia biflora* (R. & P.) F. & M.

Cardinal-flower
Lobelia cardinalis L.

[238]
Lobelia species have milky sap, alternate leaves, 2-lipped corollas, upper lip generally erect, lower lip spreading and 3-cleft. Petals united, the tube split to the base on the upper side. The 5 anthers are united. Ovary inferior. Fruit a many-seeded capsule. Many species are difficult to identify. Since few species occur in seaside habitats, identifications are easier.

The brilliant crimson flowers allow this species to be recognized at considerable distances, as other lobelias are purple to nearly white-flowered. Plants grow to 2.5 m tall. The entire flower is 23–33 mm long. Leaves are lanceolate to elliptic, to 20 cm long and 6 cm wide. Occasional. Swamps, marshes, depressions in woods, pond margins, ditches; eTX–NBr. July–Oct.

Purple Lobelia
Lobelia elongata Small

[239]
This species is a glabrous perennial to 1.6 m tall. Leaves narrowly lanceolate, tapering at both ends. Flowers are 20–25 mm long. Calyx segments entire. Corolla tube 8–14 mm long. Inner base of the lower corolla lip glabrous. Occasional. Swamps, tidal and other marshes, low ground; GA–DE. Aug.–Oct.

Only 2 other purple-flowered species are likely to be encountered in seaside habitats: *L. nuttallii* R. & S., which also has entire calyx segments but linear stem leaves. Rare. Depressions, ditches, meadows, wet woods; wFL; GA–NY. May–frost. *L. glandulosa* Walt., which has glandular calyx segments and linear to oblanceolate leaves. Rare. Similar places; FL–sNC. Sept.–Oct.

ASTERACEAE
Composite Family

Elephant's-foot
Elephantopus tomentosus L.

[240]
The Asteraceae have flowers in compact heads (See diagram on p. 44). A head may bear few to hundreds of flowers on a common receptacle and is surrounded by a few to many involucral bracts. A receptacular bract may be fastened at the base of each flower, or may be absent. Calyx represented by a pappus of scales, awns, or bristles, the latter sometimes featherlike; or a combination of these; or rarely absent. It functions in fruit transport through clinging or by wind. Petals are united. In some heads, all flowers have ligules (rays). In others, all flowers lack ligules and are called disc flowers. It also happens that ligulate and disc flowers occur in the same head. Ovary inferior, containing 1 seed, which is free from the ovary wall. Stigmas 2. Stamens are 5 and united by their anthers only. Fruit an achene or nutlet.

This species is a perennial to 60 cm tall. Leaves basal, densely hairy beneath. Flowers all disc, in heads with 3 conspicuous resin-dotted bracts at base, these 9–21 mm long. Longest involucral bracts 10–13 mm long. Corollas white to purple, 7–8 mm long. Common. Maritime and live oak woods, dry pinelands, scrub, roadsides; eFL–MD. July–Oct.

E. nudatus Gray is similar but the 3 bracts are 7–12 mm long and the longest involucral bracts 6–9 mm long. Occasional. Moist soils—pinelands, swamp margins, broadleaf woods; eFL–DE, July–Oct.

Boneset
Eupatorium perfoliatum L.

[241]
Seaside species of this genus are mostly perennials, leaves opposite or whorled or rarely alternate, flowers all disc and bisexual, achenes 5-angled, and pappus of capillary bristles bearing small barbs pointing toward the tip. We have recorded 15 species in seaside habitats. The more common and easily identified ones are included here.

This species is a perennial to 2 m tall. Stems little to widely branched, conspicuously hairy, especially below. Easily recognized by its opposite, sessile, perfoliate, or strongly clasping leaves. Flowers 9–23 per head. Common. Moist to wet places in woods, pond margins, sloughs, swales, meadows, ditches; SC–NS. July–Oct.

Dog-fennel
Eupatorium capillifolium (Lam.) Small

[242]
Perennial to 2 m tall. Stems finely hairy, at least upper portion. Upper stem leaves usually greenish, deeply and narrowly divided, the segments about 0.5 mm wide. Flowers 3–6 in each head. Common. Open places—meadows, swales, old fields, pond borders, ditches; eTX–MD. Sept.–frost.

There are 2 other very similar species. *E. compositifolium* Walt. has hairy stems and leaf segments over 1 mm wide. Flowers very fragrant. Common. Thin woods, grasslands, old fields, swales; eTX–sNC. Aug.–frost. *E. leptophyllum* DC. has glabrous stems and leaf segments about 0.5 mm wide. Rare. Pond shores, cypress-gum swamps, depressions in thin woods or open; FL–sNC. Sept.–frost.

Coastal White Snakeroot
Eupatorium aromaticum L.

[243]
Perennial with 1 to few erect finely hairy stems to 1 m tall. Leaves with petioles 15–60 mm long, blades 2–8 cm long. Involucral bracts nearly equal, only a few overlapping. Receptacle flat. Common. Dry woods, thin scrub, between old dunes; AL–MA. Aug.–Oct.

Two other seaside species with opposite similarly shaped leaves and evident petioles, usually over 1 cm long, are likely to be encountered. *E. serotinum* Michx., to 3 m tall, has petioles 10–50 mm long, leaf blades to 20 cm long, white corollas, and involucral bracts overlapping in 3 series, with outer bracts less than ½ as long as inner. Common. Fresh or brackish marshes, pond edges, sloughs, meadows; eTX–sNJ. Aug.–Nov. The other species is *E. coelestinum* L. It has bright blue to reddish purple corollas, a conic receptacle, petioles 5–25 mm long, and leaves 45–90 mm long. Rare. Moist woods, wet meadows; eTX–sNJ. July–Oct.

Broad-leaved Eupatorium
Eupatorium rotundifolium L.

[244]
Soft-hairy perennial to 1.5 m tall, stems usually solitary. Leaves sessile or nearly so, petioles rarely to 3 mm long, blades of midstem leaves about as wide as long. Involucral bracts obtuse to sharply acute. Flowers 5–7 per head. Corollas dull whitish. Quite variable and usually divided into varieties. Common. Thin wet pinelands,

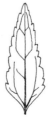

Eupatorium pilosum

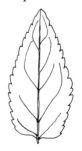

Eupatorium pilosum

Eupatorium rotundifolium

Eupatorium rotundifolium

swales, edge of sloughs, meadows, depressions in thin broadleaf woods; eTX–sME. Aug.–Oct.

E. *pilosum* Walt. is similar but blades of midstem leaves are nearly twice as long as wide. Rare. Similar places; SC–MA. Aug.–Oct.

White Eupatorium
Eupatorium album L.

[245]
Perennial to 1 m tall. Leaves sessile or nearly so, the larger ones mostly 15–30 mm wide and three times or more as long. Involucral bracts imbricate, the longest 8–11 mm, conspicuously white towards the end, tip long-acuminate. Flowers 5 per head. Corolla white. Variable and often divided into varieties. Common. Dry places—thin woods, pinelands, old fields; SC–NY. June–Oct.

Narrow-leaved Eupatorium
Eupatorium hyssopifolium L.

[246]
Perennial to 1.5 m tall from a short root stock. Leaves in whorls of 4, sometimes 2 or 3, mostly linear, 6–40 times as long as wide, spreading or ascending, clusters of smaller leaves often in the axils. Involucral bracts 4–7 mm long. Flowers 5 per head. Corolla white. Common. Dry areas— thin pinelands, thin scrub, low stable dune areas, old fields; SC–NY. July–Oct.

E. *anomalum* Nash has a similar aspect but leaves mostly 10–12 mm wide and plants with a conspicuous tuberous-thickened short rhizome. Occasional. Usually wet places— swales, meadows, wet pinelands, pond margins; eFL–sNC. July–Oct. E.

mohrii Greene is also similar but leaves mostly 3–10 mm wide and tending to be recurved or deflexed. Rare. Similar places; LA–sVA. July–Oct. Syn.: E. *recurvans* Small.

Joe-pye-weed
Eupatorium dubium Willd. ex Poir.

[247]
Perennial to 1.5 m tall. Stems speckled or covered with purple. Leaves mostly in whorls of 3–4, thick and firm, midstem ones ovate to lanceolate-ovate and with 3 prominent nerves. Involucre 6.5–9 mm high, often purplish. Corolla purple. Occasional. Freshwater marshes, low meadows, pond shores, sloughs; SC–NS. July–Oct.

Climbing Hempweed
Mikania scandens (L.) Willd.

[248]
Twining perennial, often forming masses over other low vegetation. Leaves opposite, petiolate. Flowers in each head 4, none with ligules. Corollas nearly white to lilac or light pink. Common. In open—freshwater marshes, swales, sloughs, pond margins; eTX–sME. July–Oct.

Blazing-star
Liatris graminifolia (Willd.) Willd.

[249]
Erect perennial to 1.2 m tall from a tuberous underground base. Leaves alternate, linear, usually under 7 mm wide, to 30 cm long. Flower heads longer than broad, in a spike or spikelike raceme, rarely branched. Surrounding bracts obtuse, thin, not keeled. Flowers 7–14 per head. Corolla tube hairy within towards the base. Pappus a ring of strongly barbed bristles nearly as long as the corolla. Occasional. Dry places—thin live-oak or pine woods; AL–NY. Sept.–Oct.

The only other species likely to be encountered is *L. elegans* (Walt.) Michx., which also has linear leaves, to 13 mm wide. Flower heads are narrow, contain 4–5 flowers, are numerous in a long inflorescence. Bracts surrounding the heads all slender, the inner ones with prolonged, slightly expanded, petal-like, white to pink tips. Pappus featherlike. Rare. Similar places; eTX–SC. Sept.–Oct.

Deer-tongue; Vanilla-plant
Carphephorus odoratissimus (J. F. Gmel.) Hebert

[250]
Glabrous perennial to 1.8 m tall, with a distinct odor of vanilla that sometimes can be detected from a distance. Leaves alternate, basal ones to 50 cm long and 10 cm wide. Flower heads under 1 cm long, bracts in several overlapping series. Flowers all bisexual, rays absent. Pappus of tawny to purplish finely barbed bristles 3–4 mm long. Tons of leaves are collected annually from the wild and sold for flavoring smoking tobacco. Occasional. Thin pinelands, thin mixed woods, savannas; LA–sNC. July–Oct. Syn.: *Trilisa odoratissima* (Walt. ex J. F. Gmel.) Cass.

Woolly Golden-aster
Chrysopsis gossypina Nutt.

[251]
In members of this genus the stem leaves, and often the basal ones, are alternate. Receptacular bracts are absent. Involucral bracts much overlapping. Corollas all yellow, the outer flowers of the head with rays. Pappus present in all flowers, of capillary bristles in 2 rings, the outer much shorter than the inner.

This species is a biennial or short-lived perennial to 80 cm tall, usually with several basal decumbent branches. Woolly hairs present from the plant base to the tops of the peduncles and usually extending onto the involucral bracts. Leaves numerous, sessile or nearly so, oblanceolate to oblong-elliptic, the largest about 6 cm long, tip rounded to obtuse. Rays on larger heads about 34. Occasional. Dunes, sandy pinelands; AL–sVA. Sept.–Oct. Syn.: *Heterotheca gossypina* (Michx.) Shinners.

Grass-leaved Golden-aster
Chrysopsis graminifolia (Michx.) Ell.

[252]
Perennial to 80 cm tall, often with stolonlike rhizomes, with silvery-silky appressed hairs, at least on lower portions. Leaves usually toward the base, grasslike, erect to ascending, parallel-veined, to 35 cm long. Fruits are linear achenes. Variable in amount of hairs, glands, flowers per head and number of rays. Also varies in length of bracts and corollas. Usually divided into varieties, by some into species. Common. Usually dry places—stable dune areas; grasslands; thin pinelands, scrub, and broadleaf woods; eTX–DE. June–Oct. Syn.: *C. nervosa* (Willd.) Fern.; *Pityopsis microcephala* Small; *Pityopsis aspera* (Shuttlew.) Small.

Northern Golden-aster
Chrysopsis falcata (Pursh) Ell.

[253]
Perennial with mostly decumbent stems to 40 cm long. Leaves sessile, linear, divergent, usually curved, often folded. Occasional. Dry sandy soils in open or thin woody vegetation, dunes; NJ–MA. July–Oct.

Camphorweed
Heterotheca subaxillaris (Lam.) Britt. & Rusby

[254]
This genus has characteristics similar to those of *Chrysopsis* except that in the ray flowers the pappus is lacking or exceedingly small.

Plants of this species are taprooted, annual or biennial, glandular and sticky, with a camphorlike odor when crushed. Stems to 2.5 m tall, erect, ascending, or decumbent, the latter type usually encountered in seaside habitats. Rays mostly 20–45. Common. Primary and other dunes, disturbed soils, roadsides, thin woods, old fields; TX–NY. July–Oct.

Seaside Goldenrod
Solidago sempervirens L.

[255]
Placing some plants in *Solidago* with certainty can be difficult because distinguishing characters are troublesome. Identification to species can be even more difficult because there are many species, many are quite variable, hybrids are involved, and differences depend on characters difficult to interpret. Usually goldenrods may be recognized as follows. They are perennials with alternate simple and entire or variously toothed leaves. Involucral bracts overlapping and of several lengths. Both ray and disc flowers present. Pappus a single circle of bristles. The style has flattened branches that are glabrous inside and finely hairy outside. We have recorded 17 species in seaside habitats. Most are uncommon to rare. The more common ones are included.

This species may grow to 1.8 m tall, lacks slender rhizomes, and has fleshy entire smooth leaves that are tapered at base. Rays 7–17, disc flowers 10–22. Common. Edge of salt or brackish marshes, bay shores, swales, minidunes, overwash areas; eTX–Nfld. Aug.–Nov.

Sweet Goldenrod
Solidago odora Ait.

[256]
Plants to 1.6 m tall, with a coarse rootstock. Stem leaves abundant, sessile, narrow, entire, anise-scented when bruised, those of midstem longest. Common. Thin dry woods, roadsides, thin scrub, dry areas between stable dunes; eTX–NH. July–Oct.

S. tortifolia Ell. is similar but plants have elongate rhizomes. Leaves not anise-scented and some or most leaves finely serrate. Common. Similar places; SC–sVA. Sept.–Oct.

Tall Goldenrod
Solidago canadensis L.

[257]
Plants to 2.5 m tall. Stem hairy, not glaucous. Leaves numerous, crowded along stem, those of midstem the longest, sessile, triple-nerved, finely hairy across underside. Rays 10–17. Occasional. Usually moist places— thin woods, behind established dunes, scattered shrub areas, meadows; eTX–AL; SC–Nfld. Sept.–Oct.

S. gigantea Ait. is quite similar but the stems are glabrous and glaucous. Leaves glabrous, or with hairs on underside of leaf usually confined to the 3 main veins. Occasional. Similar places; NY–NS. July–Sept.

Rough-leaved Goldenrod
Solidago rugosa Mill.

[258]
Plants to 2.5 m tall, from elongate rhizomes. Stems with spreading hairs. Leaves hairy, at least below, nearly sessile, with prominent teeth, not 3-nerved but lateral veins several from midrib. Rays 6–11, disc flowers 4–8. Common. Thin woods, moist to wet shrubby areas, meadows, swales, freshwater marshes, bog edges; NC–Nfld. Aug.–Oct.

S. fistulosa Mill. has the same general appearance but the leaves are sessile, clasp the stem, sometimes barely so, and have less conspicuous teeth. Occasional. Similar habitats; SC–NJ. Aug.–frost.

Goldenrod
Solidago uliginosa Nutt.

[259]
Plants to 1.5 m tall, the one illustrated unusually small. Stem glabrous below inflorescence. Leaves on lower third of plant much larger than those at midstem, upper surface smooth, blades gradually tapered at base. Branches and axis of inflorescence finely hairy. Achenes glabrous. Occasional. Usually in moist to wet places—bogs, meadows, thin scrub areas, ditches; NY–Nfld. July–Oct.

S. nemoralis Ait. also has prominent basal leaves but is smaller, to 1 m tall, has minute spreading hairs on leaves and stem, and leaves rough above. Inflorescence long and narrow to as broad as long. Ray flowers 5–9, disc flowers 3–8. Common. Usually dry places—thin woods, meadows, between stable dunes, roadsides; nNC–NS. July–Oct.

Flat-topped Goldenrod
Euthamia tenuifolia (Pursh) Greene

[260]
Euthamia species have many features of *Solidago* and have been included in that genus by many people. The distinctive flat-topped inflorescences and fine glandular dots on the leaves distinguish *Euthamia*. Only *S. odora* has such glands.

E. *tenuifolia* has 1-nerved leaves 1–4 mm wide and 20–50 times as long as wide. Ray flowers 7–16, disc flowers 3–9. Common. Brackish and freshwater marshes, roadsides, swales, slough margins, meadows, thin woods, overwash areas; eLA–ME. Aug.–Oct. Syn.: *E. minor* (Michx.) Greene.

E. *graminifolia* (L.) Nutt. is similar but leaves mostly 3-nerved, 3–10 mm wide, 10–20 times as long as wide, ray flowers 15–35, disc flowers 4–13. Common. Similar places; VA–Nfld. Aug.–Sept.

Lazy Daisy
Aphanostephus skirrhobasis (DC.) Trel. var. *thalassius* Shinners

[261]
Annual to 50 cm tall, erect to spreading, along the coast mostly the latter. Pappus an irregular scaly crown or of separate scales, under 1 mm long. Achenes columnar, lacking ribs, 4-sided, grooved. Occasional. Beaches, dunes, sandy meadows; TX–wFL. Mar.–June.

Perennial Saltmarsh Aster
Aster tenuifolius L.

[262]
We have recorded 23 species of *Aster* in seaside habitats. Most are difficult to name. Asters have characters in common with *Solidago* and *Erigeron*. Species of the former genus may be separated by their yellow rays; those of *Erigeron* and *Aster* are blue to purple, reddish, pink, or white. Time of flowering is helpful in separating the latter two; most species of *Erigeron* bloom in the spring to early summer and those of *Aster* usually in late summer to fall. We are omitting some of the rare species.

This species is a perennial to 70 cm tall from slender creeping rhizomes. Stems appear somewhat dichotomously branched, usually zigzagging. Leaves few, fleshy, linear or nearly so, the lower falling early. Flower heads few to many, scattered; involucre 6–9 mm high. Rays 15–25, blue to pink or nearly white, 4–7 mm long. Common. Salt and brackish marshes, sand-mud flats; eTX–MA. June–Dec.

Annual Saltmarsh Aster
Aster subulatus Michx.

[263]
Glabrous annual to 1.5 m tall. Main stem straight or nearly so. Leaves entire, linear. Flower heads few to many but well separated, involucre 5–8 mm high. Rays 15–50, bluish to sometimes white, about 3 mm long. Common. Salt and brackish marshes, mud flats, brackish swales and sloughs, ditches; TX–sME. July–Nov.

New York Aster
Aster novi-belgii L.

[264]
Slender to stout perennial to 1.4 m tall. Leaves lanceolate to elliptic, sessile, auriculate-clasping, sharply serrate to entire, glabrous except for scabrous-ciliate margins, often thick and firm. Flower heads several to many, involucre glabrous, 5–10 mm high. Rays 20–50, blue or occasionally white to rose, 6–14 mm long. Common. Moist places—salt and brackish marshes, meadows, swamp borders, pond shores, slou)ws; SC–Nfld. July–Oct.

A. *novae-angliae* L. (New England Aster) is similar but the upper leaf surface is scabrous or stiffly appressed-hairy and the involucres or peduncles or both are glandular. Rays 45–100, bright reddish-purple or rosy.

Occasional. Moist places—meadows, pond shores, sloughs, thin woods, low shrub areas; NY–NS. Aug.–Oct. *A. patens* Ait. is also similar but leaves are cordate-clasping, hairy or scabrous above, involucres barely glandular to glandular or short hairy or both. Rays 15–30, blue or rarely pink. Rare. Usually dry places—thin woods, old fields, meadows; eTX–MA. Aug.–Oct.

Wood Aster
Aster divaricatus L.

[265]
Perennial to 1 m tall. Leaves glabrous or with some long appressed hairs; lower leaves ovate with cordate base, acuminate; the lowest often smaller than those above and often dying early; the upper ones progressively less cordate, less petiolated, and smaller. Involucre 5–10 mm high, the outer bracts about 2.5 times or less as long as wide. Rays 5–16, white. Rare. Dry places—thin woods; NY–sME. Aug.–Oct.

A. undulatus L. (Wavy-leaf Aster) is the only other seaside species likely to be encountered that has cordate leaves. Flower heads in a panicle that also contains numerous small bracts. Involucre bracts three times or more as long as wide. Rays 10–20, blue to lilac. Occasional. Similar places. SC–NS. Aug.–Nov.

Many-flowered Aster
Aster lateriflorus (L.) Britt.

[266]
Perennial to 1.2 m tall. Midstem leaves entire to serrate, sessile or nearly so, linear to lanceolate or nearly rhombic, 5–15 cm long and 5–30 mm wide. Flower heads usually many, small, involucre 4.5–5 mm long, involucral bracts acute to obtuse and glabrous. Rays 9–14, white to slightly purplish. Disc corollas deeply lobed, the lobes composing 50–95% of the expanded portion. Common. Usually in dry places—thin woods, fields, meadows, sloughs, pond shores; nSC–NS. Aug.–Nov.

Aster dumosus

Two other species have small flower heads and are otherwise of similar aspect: *A. dumosus* L., which differs mainly in having lobes of disc corollas composing 20–35% of the expanded portion. Common. Thin woods, fields, meadows, freshwater marshes, thin pinelands, pond borders; eTX–sMA. Aug.–Nov. *A. ericoides* L. (Heath Aster), which differs mainly in having some or all involucral bracts with coarsely ciliate margins. Occasional. Dry places—thin woods or in open; MD–sME. July–Oct.

Subulate-bracted Aster
Aster pilosus Willd.

[267]
Perennial to 1.5 m tall. Stems and leaves glabrous to hairy. Midstem leaves linear to lanceolate-elliptic, sessile or nearly so, about 10 cm long and 1–2 cm wide. Heads few to many, fairly well spaced. Involucres 3.5–8 mm high, broadly urn-shaped. Involucral bracts with tips glabrous, green, spreading, subulate, and marginally inrolled. Flowers 40–100 per head, rays 16–35, usually white, uncommonly pink to lavender or purple. Common. Thin woods and scrub, old fields, meadows, edges of freshwater and brackish marshes, roadsides; SC–NS. Aug.–Nov.

Linear-leaved Aster
Aster linariifolius L.

[268]
Perennial to 60 cm tall with 1 to several wiry stems from the base. Leaves numerous, similar, firm, linear or nearly so, entire. Heads solitary or more often a few to many in a fairly compact cluster. Involucre 6–9 mm

high. Rays 10–20, violet or rarely white. Pappus double—the inner ones tawny, long, capillary; the outer ones short bristles about 1 mm long. Common. Dry open places, thin woods, usually sandy or rocky soils; eTX–wFL; SC–ME. Aug.–Nov.

White-topped Aster
Aster paternus Cronq.

[269]
Perennial to 60 cm tall. Leaves ciliate-margined, some toothed; the basal and lower stem leaves generally the largest, persistent, broadly oblanceolate to obovate or elliptic, tapering to petioled bases. Leaves near the top sessile. Flower heads in corymblike clusters. Involucres 5–9 mm high, glabrous. Rays 4–8, white to rarely pink, 4–8 mm long. Disc flowers 9–20. Occasional. Thin woods, roadbanks, stable dune areas; nSC–sME. June–Aug. Syn.: *Sericocarpus asteroides* (L.) B.S.P.

Oak-leaf Erigeron
Erigeron quercifolius Lam.

[270]
Erigerons are much like asters, but generally flower in the spring and early summer whereas most asters flower in late summer to fall. Also, involucral bracts lack green tips, a characteristic of most asters. Erigerons are also similar to *Conyza*, but the rays are over 3 mm long whereas those of *Conyza* are about 1 mm long or less.

This species is an erect biennial or short-lived perennial to 60 cm tall. Leaves are mostly basal, conspicuously hairy, clasping the stem, and usually shallowly lobed. The involucre is 2.5–4 mm high and sticky-hairy. Rays 100–250, 5 mm long. Disc corollas 1.5–2.5 mm long. Common. Waste places, thin woods, stable dune areas, old fields, roadsides; LA–NC. Mar.–June.

Robin's-plantain
Erigeron vernus (L.) T. & G.

[271]
Perennial to 60 cm tall. Leaves gla-

brous to sparsely appressed-hairy, nonclasping; basal ones prominent, oblanceolate to suborbicular, coarsely toothed. Stem leaves few. Involucre 3–4 mm high, glabrous or sparsely hairy. Rays 25–40, 4–8 mm long. Disc flowers 2.5–3.8 mm long. Occasional. Moist places—thin pinelands, pond margins, ditches; LA–sVA. Mar.–July.

Trailing Erigeron
Erigeron myrionactis Small

[272]
Hairy perennial with trailing stems to 1 m long, often rooting at the nodes. Leaves obovate to spatulate, 2–8 cm long, coarsely few-toothed. Heads solitary, on erect peduncles to 20 cm long. Rays very numerous, white, 5–7 mm long. Common. Sandy soils—open areas among and behind stable dunes, roadsides; TX–MS. Feb.–June; Oct.–Nov.

Erigeron myrionactis

Daisy Fleabane
Erigeron strigosus Muhl. Ex Willd.

[273]
Annual or biennial to 90 cm tall, at least the midstem with short appressed hairs. Stem leaves not crowded, sessile, not clasping, linear to lanceolate, entire to somewhat serrate, to 15 cm long and usually less than 1 cm wide. Basal leaves larger. Rays 50–100, to 6 mm long, white or rarely bluish-tinged. Common. Roadsides, old fields, waste places, thin disturbed woods; eTX–NS. Apr.–June, with a few plants blooming to Oct.

E. annuus (L.) Pers. has a similar as-

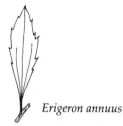

Erigeron annuus

pect but more robust, to 1.5 m tall, the stem leaves more crowded, to 7 cm wide, coarsely toothed. Stems and leaves with conspicuous spreading hairs. Occasional. Similar places but usually richer soils; nNC–NS. Apr.–June, rarely to Oct.

Horseweed
Conyza canadensis (L.) Cronq.

[274]
Conyza species resemble some *Erigeron* species but have short rays, about 1 mm long or less, and many disc flowers, whereas the latter have rays over 3 mm long and few disc flowers, about 20 or less. Seaside species are annuals.

This species is erect, to 1.5 m tall, and unbranched to branched at the base and spreading. Seaside plants are mostly var. *pusillus* (Nutt.) Cronq., which is shown in the photograph. Plants of this variety are glabrous or nearly so, frequently spreading, have some or all of the involucral bracts minutely purple-tipped. Involucres are 3–4 mm high. Common. Stable or somewhat active dunes, grasslands, meadows, thin woods, roadsides, old fields; var. *pusillus* occurs TX–CT. June–frost. Syn.: *Erigeron pusillus* Nutt.

C. canadensis var. *canadensis* is spreading, hairy, usually erect, and involucral bracts lack purple tips. Occasional. Mostly absent from dune areas but otherwise similar places; VA–NS. July–frost. Syn.: *Erigeron canadensis* L.

Hairy Fleabane
Conyza bonariensis (L.) Cronq.

[275]
Stems erect, usually only one, to 1 m tall, abundantly and loosely hairy. Involucres 4–6 mm high, with abundant short hairs. Rays minute, to 0.5 mm long. Flowers to 200, occasionally more, in each head. Occasional. Waste places, roadsides, old fields; eTX–sNC. Apr.–Oct. Syn.: *Erigeron bonariensis* L.

Marsh-fleabane
Pluchea odorata (L.) Cass.

[276]
Members of this genus are strongly aromatic, have alternate leaves, involucral bracts of several lengths and overlapping, no rays, no receptacular bracts, and a pappus of a single ring of fine bristles.

This species is an annual to 1.5 m tall. Leaves petioled or sometimes the upper tapered to a narrow base. Central clusters of flower heads exceeded by some of the lateral flowering branches. Involucral bracts usually pink to purple, at least on the tips, bearing small several-celled glandular-sticky hairs. Corollas rose-purplish. Common. Salt and brackish marshes, sloughs, and swales; salt flats; rarely in freshwater marshes; TX–MA. Aug.–Oct. Syn.: *P. purpurascens* (Sw.) DC.

The only other species with petioled leaves in seaside habitats is *P. camphorata* (L.) DC. in which the central clusters of flower heads overtop the lateral clusters and the involucral bracts are merely finely-glandular, lacking several-celled hairs. Occasional. Freshwater habitats—marsh edges, meadows, swales, swamps; TX–DE. Aug.–Oct.

Stinkweed
Pluchea foetida (L.) DC.

[277]
Perennial to 1 m tall, glandular and often somewhat cobwebby and short hairy. Leaves sessile, oblong-elliptic to ovate, usually broad-based, and clasping stem. Heads several to many in a short broad flat-topped cluster or sometimes storied. Corollas creamy-white. Common. Wet soils—mead-

ows, pond edges, swampy woods, edge of freshwater marshes; eTX–MD. July–Oct.

P. rosea Godfrey is quite similar but corollas are rose-pink to rose-purple. Common. In woods or open—wet pinelands, pond shores, intermittent ponds, swales, ditches, poorly drained woods; eTX–sNC. June–July.

Black-root
Pterocaulon pycnostachyum (Michx.) Ell.

[278]
Perennial to 80 cm tall, from large dark roots. Stem conspicuously winged from decurrent leaf bases. Leaf undersides and stem with the feel of kid leather due to short densely felted hairs; hairs light-colored. Flowers very small, in small compact heads, these in tight elongate clusters. Corollas yellow. Common. Dry to moist places—thin pinelands, swales; eFL–sNC. May–June.

Rabbit-tobacco
Gnaphalium chilense Spreng.

[279]
Members of this genus are annuals or biennials. Flower heads are sessile in various types of clusters. Involucral bracts thin, dry, and membranous towards the tip, woolly at the base. Pappus of capillary bristles. Receptacular bracts absent.

This species is an erect annual or biennial to 80 cm tall. Leaves numerous and narrow, sessile, decurrent on stem, most diverging less than 45°. Flower heads in obviously terminal clusters. Involucres 4–6 mm high. Pappus bristles distinct and falling separately except sometimes cohering in small groups by means of tiny interlocking basal hairs. Occasional. In open well-drained sandy places—old fields, grasslands, between old dunes, roadsides; nSC–VA. May–Aug.

Rabbit-tobacco; Everlasting
Gnaphalium obtusifolium L.

[280]
Erect fragrant annual or winter-

annual to 1 m tall. Similar to *G. chilense* except leaves are not decurrent, are spreading at right angles or nearly so, and are glabrous to only slightly woolly above. Common. Thin woods, thin pinelands, old fields, stable dune areas, roadsides; eTX–NS. Aug.–Nov.

Anaphalis margaritacea (L.) Benth. & Hook. (Pearly-everlasting) has a similar aspect but is a rhizomatous perennial; the flower heads, or most of them, are individually short-pedunculate and larger, to nearly 1 cm broad. Rare. Thin woods or open, usually dry places; NY–Nfld. July–Sept.

Purple Cudweed
Gnaphalium purpureum L. var. *purpureum*

[281]
Annual or biennial to 60 cm tall. Leaves on lower half of plant mostly oblanceolate to spatulate, obviously greener and less hairy on the upper surface than on the lower. Heads numerous in a terminal somewhat leafy-bracted cluster. Involucre 3–5 mm high, bracts mostly acute to acuminate, light brown, and usually tinged with purple. Pappus falling as a unit, the bristles being united into a ring at the base. Common. Around buildings, roadsides, old fields, thin woods, pond margins; eTX–ME. Mar.–July.

G. purpureum var. *falcatum* (T. & G.) Lam. is similar but leaves are linear to linear-oblanceolate and equally hairy on both sides. Occasional. Similar places; eTX–MD. Mar.–July. *G. uliginosum* L. has an aspect similar to the latter variety but is usually more branched, the leaves often shorter, the clusters of flower heads conspicuously overtopped by the intermixed leaves, and the pappus bristles falling separately. Rare. Similar places; NY–Greenl. July–Oct.

Common Ragweed
Ambrosia artemisiifolia L.

[282]
Taprooted annual to 2 m tall. Leaves

deeply dissected, the lower ones opposite, upper alternate. Flowers unisexual, male and female in separate heads, male flowers in spikelike racemes, female just below these in axils of small leaves. Female flowers and fruits surrounded by a hard angular structure bearing 5–6 sharp erect spines near the apex. Common. Roadsides, old fields, waste places, pond and marsh margins, northern sand and gravel beaches; TX–Nfld. July–frost.

Cocklebur
Xanthium strumarium L.

[283]
Coarse freely branched taprooted annual to 2 m tall. Leaves broadly ovate to suborbicular and usually cordate or nearly so at base. Flowers unisexual, the male in many-flowered heads at ends of branches, dying and usually falling off before the fruits fall (as in the picture). Female flowers, 2 each in burs beset with strong hooked bristles. Common. Waste places, cultivated and old fields and gardens, pond shores, ditches, stable dune areas, beaches; TX–Nfld. July–frost. Syn.: *X. echinatum* Murray.

Eclipta
Eclipta prostrata (L.) L.

[284]
Little to much-branched annual bearing scattered stiff appressed hairs pointing upward. Stems to 1 m long, branches weakly ascending to prostrate, often rooting at nodes. Leaves opposite, linear-lanceolate to lanceolate or lance-elliptic. Flower heads occur 1–3 in axil of one of a pair of leaves. Rays white, 1–2 mm long. Disc corollas dusky-white. Receptacular bracts bristlelike. Pappus absent or at most an inconspicuous crown. Achenes thick, truncate at apex, 4-angled or some outer ones 3-angled. Common. Pond shores, sloughs, ditches, freshwater marshes, depressions; TX–MA. June–frost. Syn.: *E. alba* (L.) Hassk.

Black-eyed-Susan
Rudbeckia hirta L.

[285]
Taprooted annual to a fibrous-rooted perennial, to 3 m tall, with abundant coarse spreading hairs. Leaves alternate, elliptic to ovate, entire to coarsely toothed, attached along the stem. Involucral bracts quite hairy. Receptacle well rounded. Rays yellow, disc corollas blackish purple. Disc flowers fertile and accompanied by receptacular bracts; ray flowers sterile and bractless. Pappus absent. Achenes glabrous, quadrangular. Occasional. Dry places—thin woods, roadsides, old fields, scattered scrub, grasslands; SC–NS. May–frost. Syn.: *R. serotina* Nutt.

Narrow-leaved Sunflower
Helianthus angustifolius L.

[286]
Erect perennial to 2 m tall with a single stem that often has many branches in the upper half. Leaves mostly alternate, to 20 cm long, occurring all along the stem, mostly 10–30 times as long as wide. Involucral bracts several and overlapping, lanceolate to lanceolate-linear, 1–2.5 mm wide. Rays yellow, the flowers sterile. Disc corollas reddish-purple, the flowers fertile. Achenes nonwinged, pappus of 2 awnlike scales. Common. Moist places—freshwater marshes, pond margins, pinelands, ditches, meadows, slough margins; eTX–AL; eFL–NY. July–frost.

Cucumber-leaved Sunflower
Helianthus debilis Nutt.

[287]
Annual or short-lived perennial with prostrate to erect stems to 1 m tall. Leaves scabrous or rough-hairy to nearly glabrous, bases truncate to cordate. Flower heads on naked peduncles. Characteristics of the flowers and fruits are quite similar to those of *H. angustifolius*. Occasional. Beaches, dunes, waste places, dry grasslands; TX–FL: rare, GA–ME. May–Oct.

Melanthera
Melanthera nivea (L.) Small

[288]
Coarse erect rough-hairy perennial to
2 m tall. Stem 4-angled. Leaves petioled, opposite except uppermost,
toothed, blades narrow to broadly
ovate or triangular, unlobed or lobed
at base. Flower heads 1–2 cm wide,
rays absent, disc corollas white, anthers black with white tips. Achenes
4-angled, warty, apex truncate with 4
awns 2–3 mm long. Occasional. Moist
to dry live oak woods, pinelands,
beaches, dune areas; LA–sSC. June–
frost. Syn.: *M. hastata* Michx.

Crown-beard; Wingstem
Verbesina occidentalis (L.) Walt.

[289]
Erect leafy-stemmed perennial to 2.5
m tall. Stems 4-winged. Leaves opposite. Involucral bracts overlapping,
less than 3 mm wide. Rays 2–5, yellow, 5–20 mm long. Ray flowers fertile. Disc corollas yellow. Achenes
flattened, not winged, with 2 strong
awns. Common. Thin woods, old
fields, roadsides, meadows; AL–VA.
July–Oct.

Tickweed
Verbesina virginica L.

[290]
Perennial to 2.5 m tall with a single
densely fine-hairy winged stem.
Leaves alternate, ovate to lanceolate-
ovate or lanceolate-elliptic, light
green beneath, to 25 cm long, with
winged petioles. Rays 1–5, white, 5–
10 mm long. Disc corollas white.
Achenes flattened, hairy, usually
winged, with 2 short awns. Common.
Live oak woods, maritime woods,
meadows, thin scrub; TX–VA. July–
Oct.

Coreopsis
Coreopsis tinctoria Nutt.

[291]
Coreopsis species are rather easily distinguished from most other seaside
plants. They are unusual in that the

flower heads have 2 distinct series of
involucral bracts: the outer green,
narrower than the inner, and usually
spreading; the inner series brownish
to yellowish and erect. Also, the
achenes are marginally winged and
have no pappus, or the pappus consists of a minute crown or 2 barbless
teeth. *Bidens* species have similar involucral bracts but the achenes are
not winged. The pappus consists of
2–4 teeth or awns that are retrorsely
or antrorsely barbed, or barbless; or
pappus rarely absent. Both genera
have opposite basal and midstem
leaves.

This species is a glabrous or nearly
so annual to 1.5 m tall, erect to
branching at the base and spreading.
Lower and midstem leaves once- or
twice-pinnately divided, the segments narrowly linear to linear-
lanceolate or oblanceolate. Rays yellow with a reddish-brown base. Disc
corollas yellow, 4-lobed. Pappus minute to absent. Occasional. Meadows,
depressions, ditches; TX–AL; nNC–
sVA. Apr.–July. Syn.: *C. cardaminae-
folia* (DC.) Nutt.

C. lanceolata L. is the only other Coreopsis likely to be encountered in
seaside habitats. Rays are completely
yellow, the disc corollas are 5-lobed,
and leaves occur mostly toward the
base. The pappus of 2 short chaffy
teeth. Occasional. Dry places—thin
woods, roadsides, waste places,
around buildings; eTX–MA. May–
July.

Wild-goldenglow; Bur-marigold
Bidens laevis (L.) B.S.P.

[292]
In *Bidens* species, as in *Coreopsis*, the
involucral bracts are in 2 series; the
outer distinctly green and often enlarged, sometimes leafy; the inner series membranous, often lined. For
other differences and similarities see
under *C. tinctoria*.

This species is a glabrous annual to
1.5 m tall. Leaves are simple, sessile,
sometimes tapering to a narrow base,
coarsely serrate to nearly entire.
Heads many-flowered, showy, rays

15–30 mm long. Achenes 5–8 mm long, pappus of 4 awns. Common. Wet places, often in shallow water—sloughs, ditches, freshwater and brackish marshes, pond margins; eTX–NH. Aug.–Nov.

Two other seaside species have showy flower heads. They resemble each other in being glabrous, having petioled leaves that are mostly pinnately dissected or pinnately compound, and with rays 10–25 mm long. One species, *B. mitis* (Michx.) Sherff, may be recognized by achenes 1.5–2 times as long as wide with pappus absent or consisting of very short teeth or of 2 triangular finely antrorsely barbed teeth. Common. Similar places; eTX–MD. Aug.–Oct. The other species, *B. coronata* (L.) Britt., has achenes 2.5–5 times as long as wide and a pappus consisting of 2 short strong bristles or awned scales. Occasional. Freshwater or brackish marshes, peaty meadows, swales; nGA–MA. Aug.–Oct.

Bidens mitis   *Bidens mitis*

Spanish-needles
Bidens bipinnata L.

[293]
The presence of this species is probably realized most often by its needle-shaped fruits sticking to people's clothing. The fruits are linear, 4-sided, mostly 10–13 mm long, and have 3–4 yellowish retrorsely barbed awns. Flower heads are narrow and erect. Rays yellow, inconspicuous, 1 to few or occasionally absent. Plants are annual, to 1.7 m tall but will flower when quite short. Leaves opposite except sometimes the uppermost, 2–3 times pinnately dissected. Common. In woods, around buildings, stable dune areas, gardens, fields, old fields, waste places; eTX–RI. July–Oct.

Shepherd's-needle
Bidens pilosa L.

[294]
Annual to 2 m tall. Stems much branched, sometimes spreading and rooting at the nodes. Leaves largely pinnately 3- to 5-parted or -compound. Rays white, usually 5, to 15 mm long. Achenes linear but, unlike *B. bipinnata*, flat to 4-sided and compressed one way. Pappus of 2–3 yellowish retrorsely barbed awns. Occasional. Disturbed places, around buildings, old fields; eTX–sNC. Frost-free periods.

Beggar-ticks
Bidens frondosa L.

[295]
Annual to 1.2 m tall. Leaves pinnately compound with 3–5 lanceolate, acuminate, and serrate leaflets to 10 cm long and 3 cm wide. Outer involucral bracts 5–10, green, usually longer than remainder of flower head. Unlike previous *Bidens* species in having rays absent or when rarely present only to 3.5 mm long, and having flat wedge-shaped achenes 5–10 mm long with a prominent vein on each face. Disc corollas are yellow. Common. Moist to wet places—woods, edges of ponds and brackish marshes, freshwater marshes, ditches, gardens, sloughs; eTX–Nfld. Aug.–Oct.

Flower heads of *B. tripartita* L., an annual to 2 m tall, are similar, especially in that rays are absent, but the leaves are simple and serrate or uncommonly 3–7 cleft. Occasional. Freshwater marshes, swamps, wet meadows, waste places, low fields, pond edges; nVA–NS. Aug.–Oct.

Bidens tripartita

Bitterweed
Helenium amarum (Raf.) H. Rock

[296]
Helenium species have alternate leaves, usually yellow rays and disc corollas, no receptacular bracts, a pappus of 5–10 awn-tipped scales, and truncate style tips.

Bitterweed is a glabrous annual to 1 m tall, leaves linear, to 8 cm long, rarely over 2 mm wide, often with axillary clusters of smaller leaves. A serious pest in pastures. Although bitter and usually avoided by grazing animals, it is often eaten when forage is scarce. The milk produced by these animals has a bitter flavor. Common. Roadsides, dune areas, thin woods, meadows, old fields; eTX–VA. May–frost.

Sneezeweed
Helenium autumnale L.

[297]
Perennial to 2 m tall. Stem winged and leafy throughout. Leaves toothed, mostly elliptic to oblong or lanceolate, narrowed to a sessile base or nearly so. Heads several to numerous. Rays mostly 13–21, yellow, the flowers pistillate. Disc corollas yellow, 5-lobed. Pappus of scales. Reported as poisonous when eaten. Occasional. Moist to wet places—marshes, thin woods, meadows, pond margins, ditches, depressions; eFL–CT. Aug.–Nov.

Gaillardia; Fire-wheel
Gaillardia pulchella Foug.

[298]
Annual, or biennial in warmer parts of seacoast, to 70 cm tall, decumbent to erect, branches few to many, plant often dense. Leaves alternate, entire to serrate or pinnately cut. Flower heads long-peduncled, involucral bracts overlapping. Rays 6–15, red to purplish red, or the tips yellow, or completely yellow, 15–25 mm long. Disc flowers fertile, corollas brownish yellow. Receptacular "bracts" of bristles. Pappus of 6–10 awned scales. Common. Sandy soils of beaches, dunes, roadsides, grasslands, thin woods; TX–MD. Apr.–frost.

Scentless Chamomille
Matricaria maritima L.

[299]
Nearly scentless annual to 60 cm tall. Leaves alternate, to 8 cm long, twice-pinnately cleft, the segments linear to nearly filiform. Rays 12–25, white. Involucral bracts in 2–3 series but little overlapping in each series. Receptacular bracts absent. Pappus a minute crown or absent. Occasional. Waste places, around buildings, roadsides; NY–NS. July–Oct.

Anthemis arvensis L. (Dog-fennel) has a quite similar aspect but the plants are malodorus, have involucral bracts in several series and overlapping, and have chaffy receptacular bracts. Occasional. Similar places; nVA–NS. May–Aug. *Matricaria matricarioides* (Less.) Porter (Pineapple-weed) is much like both above species vegetatively and in some other aspects but rays are absent. Rare. Similar places; NJ–Greenl. June–frost.

Yarrow
Achillea millefolium L.

[300]
Aromatic rhizomatous perennial to 1.2 m tall. Leaves alternate, to 15 cm long, 2–3 times pinnately dissected, the segments linear. Involucral bracts in several series, overlapping. Ray

flowers 3–5, rays white. Disc flowers fertile. Common. Around buildings, old fields, roadsides, waste places, meadows; SC–Greenl. Apr.–frost.

Ox-eye Daisy
Leucanthemum vulgare Lam.

[301]
Perennial to 1 m tall with short rhizomes. Leaves alternate, the numerous basal ones usually pinnately lobed or cleft. Flower heads 1 or a few, with no receptacular bracts. Rays 15–35, white, 10–25 mm long. Disc flowers producing the achenes. Pappus absent. Useful as an ornamental, either in clusters or colonies, or as cut flowers, but can become a serious pest. Common. Roadsides, lawns, around buildings, waste places; NC–Greenl. Apr.–Aug. Syn.: *Chrysanthemum leucanthemum* L.

Dusty-miller; Beach Wormwood
Artemisia stelleriana Bess.

[302]
Densely white-haired perennial to 75 cm tall, with extensive creeping and branching rhizomes and decumbent leafy stems. Often forming dense colonies. Leaves alternate, bluntly lobed. Inflorescence racemelike, involucres 6–7.5 mm high, rays none, receptacular bracts absent, pappus none. Common. Beaches, dunes; VA–NS. May–Sept.

Fireweed
Erechtites hieracifolia (L.) Raf. ex DC.

[303]
Annual to 3 m tall. Stem not winged. Leaves alternate, many, not forming a basal rosette, to 20 cm long and 8 cm wide. Flower heads cylindric with a swollen base. Involucral bracts essentially equal and in one series, often with a few small ones at base. Rays absent. Outer flowers pistillate, lacking pollen-bearing anthers. Receptacular bracts absent. Pappus white, of numerous capillary bristles. Common. Dry to wet places—thin pinelands, thin woods, disturbed places,

swales, freshwater marshes, ditches, burned-over areas; TX–NS. July–frost.

Hairy Groundsel
Senecio tomentosus Michx.

[304]
Perennial to 70 cm tall, spreading by basal offshoots and stolons, cottony-hairy, especially at base, sometimes losing much of the hair after flowering. Leaves alternate, crenate to nearly entire, mostly basal. Involucral bracts in one series with a few very short ones at base. Rays yellow, 5–10 mm long. Receptacular bracts absent. Pappus nearly white, of capillary bristles. Occasional. Usually wet places in thin woods or open—between stable dunes, grasslands, meadows; VA–sNJ. Apr.–June.

Yellow Thistle
Cirsium horridulum Michx.

[305]
Seaside *Cirsium* species are easily recognized by spiny-margined leaves, involucral bracts usually spiny, absence of rays, and a pappus of numerous plumose bristles.

This species is a biennial to 1.5 m tall, leaves very spiny and not decurrent on stem, and with a series of narrow spiny-toothed leaves tightly surrounding base of the flower heads. Corolla light yellow or white to purple. Common. Dune areas, roadsides, meadows, thin woods, pinelands; TX–sME. Mar.–Aug. Syn.: *Carduus spinosissimus* Walt.

Bull Thistle
Cirsium vulgare (Savi) Ten.

[306]
Biennial to 1.5 m tall. Stem with conspicuous spiny wings. Involucre not surrounded at base by spiny leaves. Common. Roadsides, old fields, thin woods, meadows, waste places; VA–Nfld. June–Sept.

Only 2 other thistles are likely to be encountered in seaside habitats. One, *C. arvense* (L.) Scop. (Canada Thistle), can be recognized because it is a perennial. Involucres are 1–2 cm high and the bracts sharp-pointed but not spiny. Common. Similar places; nNJ–Greenl. June–Oct. The other species, *C. muticum* Michx. (Swamp Thistle), a biennial, has spineless involucral bracts, and the involucre is 20–35 mm high. Occasional. Swamps, bogs, swales, wet meadows; DE–Nfld. July–Sept.

Common Chicory; Blue-sailors
Cichorium intybus L.

[307]
Milky-juiced perennial to 1.7 m tall from a long taproot. Stem usually much branched. Basal leaves numerous. Flower heads sessile or short-peduncled, 1–3 of them in axils of the much smaller upper leaves. All flowers with rays, these bright blue or rarely pink or white. Pappus of 2–3 rows of very short scales. Occasional. Roadsides, old fields, waste places, fencerows; nNC–NS. May–Oct.

Dwarf-dandelion
Krigia virginica (L.) Willd.

[308]
In *Krigia* species all flowers have yellow rays, the involucral bracts are of equal size and the pappus absent or of scales and/or scabrous bristles. The juice is milky.

This species is an annual to 40 cm tall, leaves all basal, the involucral bracts 9–18 and 4–7 mm long, pappus of 5 short thin scales alternating with 5 much longer scabrous bristles. Common. Dune areas, roadsides, thin woods, lawns, waste places; eTX–NH. Mar.–Aug.; rarely during mild winters in south.

K. caespitosa (Raf.) Chambers may also be encountered in seaside habitats. An annual with stem as well as basal leaves, stems often several. Pappus absent or minute. Occasional. Usually moist places—roadsides, old fields, thin pinelands; TX–wFL. Mar.–June. Syn.: *K. oppositifolia* Raf.

Dandelion; Blowballs
Taraxacum officinale Weber

[309]
Perennial from a deep taproot to 1 cm thick. Leaves all basal, barely lobed to sharply pinnately cut or divided. Flowers with rays only, the heads single on each stem. Stems hollow, to 50 cm tall. Achenes are brown to olive- or straw-colored, have a long thin neck at top of which is a "parachute" composed of numerous capillary bristles. Common. Roadsides, lawns, waste places, grasslands; NC–Nfld. Mar.–Sept.

T. laevigatum (Willd.) DC. is almost identical, the leaves sometimes less deeply cut and the achenes red to purplish red or brownish red at maturity. Rare. Similar places; VA–NS. Apr.–July. Syn.: *T. erythrospermum* Andrz. ex Bess.

Fall-dandelion
Leontodon autumnalis L.

[310]
Perennial from a short rootstock. Stem to 40 cm tall, often decumbent at base. Leaves mostly basal. All flowers with yellow rays. Receptacular bracts absent. Pappus wholly of plumose bristles that are chaffy-flattened at base. Occasional. Roadsides, dune areas, meadows, lawns, waste places; NJ–Greenl. May–Nov.

Hypochoeris radicata L. (Cat's-ear) has a similar aspect but receptacular bracts are present and chaffy, and the pappus consists of shorter minutely barbed outer bristles and plumose inner ones. Occasional. Similar places; NC–Nfld. Apr.–Aug.

Prickly Sow-thistle
Sonchus asper (L.) Hill

[311]
Annual to 2 m tall. Leaves prickly, sometimes pinnately divided, the 2 basal lobes rounded but prickly-toothed. Heads with numerous flowers, over 80, all with rays, these 8–13 mm long. Pappus of numerous fine white very soft bristles. Achenes lacking a beak, somewhat flattened, with

3–5 ribs on each face. Common. Waste places, roadsides, old fields, meadows, disturbed areas; TX–Greenl. Mar.–Oct.

S. oleraceus L. (Common Sow-thistle) is very similar, but is weakly prickly, the basal lobes of the leaves distinctly acute, and the achenes transversely roughened as well as with ribs, though these are sometimes indistinct. Common. Similar places; TX–Greenl. Mar.–Oct. *S. arvensis* L. (Perennial Sow-thistle) is also similar but is a perennial with larger heads, the rays being 15–25 mm long. Occasional. Similar places; NJ–Nfld. July–Oct.

Wild Lettuce
Lactuca canadensis L.

[312]
Some *Lactuca* species are much like *Sonchus* species, but the former have only 11–55 flowers in each head and the achenes are enlarged at the summit.

This species is a leafy-stemmed annual or biennial to 2.5 m tall, with abundant sticky milky juice. Leaves scarcely prickly. Rays yellow. Achenes flattened, 4.5–6.5 mm long, with only a medial nerve on each face, and with a beak ½–1 times as long as the body. Common. Roadsides, old fields, disturbed places, thin woods, waste places; nSC–NS. June–frost.

Lactuca canadensis

Lactuca canadensis

The other species likely to be encountered in seaside habitats are as follows: *L. serriola* L. (Prickly Lettuce) with leaves prickly on the midrib below, achenes only 3–4 mm long and with several nerves and a very slender beak 1–2 times as long as the body. Occasional. Similar places; nVA–NBr. June–frost. Syn.: *L. scariola* L. Also *L. graminifolia* Michx. with mostly basal leaves, blue rays, and achenes 6–9 mm long. Rare. Thin woods, disturbed places; eFL–sNC. Apr.–July.

False-dandelion
Pyrrhopappus carolinianus (Walt.) DC.

[313]
Milky-juiced annual or short-lived perennial to 1.2 m tall from a prominent taproot. Stem leaves 0–12. Heads solitary to several. The longest involucral bracts 2-lobed or wider at the tip than just below it. Flowers in each head numerous. Rays yellow to pale cream-colored. Anthers dark. Achene body 4–6 mm long, with a filiform beak often twice as long. Pappus of capillary bristles with a ring of minute soft white reflexed hairs just beneath. Common. Roadsides, stable dune areas, meadows, old fields, lawns and borders, thin woods; eTX–DE. Feb.–June.

Rattlesnake-weed
Hieracium venosum L.

[314]
Hieraceum species are milky-juiced perennials from a fibrous-rooted rootstalk. Flower heads are 1 to numerous. All flowers are perfect and have rays. Achenes are terete or nearly so and beakless although sometimes narrowed towards the summit. Pappus of capillary bristles.

This species has prominent purple-veined leaves, these all or mostly basal, stem leaves to 3, rarely more. Heads several and widely spaced. Rays yellow. Common. Dry thin woods, clearings, wood borders; VA–NH. May–Sept.

Orange Hawkweed
Hieracium aurantiacum L.

[315]
Plants to 60 cm tall, leaves mostly
basal, heads mostly 5–10 in a compact
cluster, involucral bracts with gland-
tipped hairs, and rays red-orange.
Easily recognized, as no other seaside
hawkweed has flowers this color.
Common. Roadsides, lawns, old
fields, dune areas, grasslands; nNJ–
Nfld. June–Sept.
  H. pilosella L. (Mouse-ear) is also
easily recognized. It has basal leaves,
perhaps 1 stem leaf; 1 flower head,
seldom 2, rarely to 4; and yellow rays.
Rare. Similar places; NY–Nfld. June–
Sept.

King-devil
Hieracium caespitosum Dum.

[316]
Stems 1 to several, to 90 cm tall, not
glaucous, at least the lower part
hairy. Leaves mostly or all basal,
mostly 5–12 times as long as wide,
conspicuously hairy on both sides.
Stem leaves 0–3. Heads 5–30 in a
compact cluster. Rays yellow.
Achenes 1.5–2 mm long. Occasional.
Roadsides, lawns, meadows, grasslands, between stable dunes, thin
woods; nNJ–NS. May–Aug. Syn.: *H.
pratense* Tausch.
  H. piloselloides Vill. is quite similar
but is glaucous and the leaves are
sparsely hairy to glabrous. Common.
Similar places; NY–Nfld. May–Aug.
Syn.: *H. florentinum* All.

Leafy Hawkweed
Hieracium gronovii L.

[317]
Plants to 1.5 m tall. Stems mostly solitary, conspicuously long-hairy
towards base. Leaves chiefly below
the middle or towards the base. Inflorescence elongate and open-cylindric
unless plants are small. Involucre 6–9
mm high. Flowers 20–40 per head.
Common. Thin dry woods, old fields,
thin scrub; AL–MA. July–frost.
  H. megacephalum Nash is similar but
the inflorescence is short, broad, and

open; and the involucres 8–11 mm
high. Rare. Thin pinelands, thin live
oak woods, among palmettos; eFL–
GA. July–frost.

Herbaceous Monocots
with Sepals and Petals

JUNCAGINACEAE
Arrow-grass Family

Arrow-grass
Triglochin striata R. & P.

[318]
Arrow-grasses are perennials with all
leaves basal, narrowly long-linear,
fleshy, terete, their bases sheathing
the stem. Flowers perfect, many, on
short pedicels in bractless racemes.
Perianth segments 3 or 6. Ovary superior, of 3, 4, or 6 weakly united carpels that separate and fall at maturity.
  This species grows to 35 cm tall,
has 3 perianth segments and 3 carpels. Fruit nearly globose, about 3
mm high. Common but usually inconspicuous. Salt or brackish
marshes; brackish water in ditches,
sloughs, ponds, overwash areas; LA–
MD. May–Oct.
  Two other species may be encountered in seaside habitats. They are
much like the above species. *T. maritima* L. is often more robust, growing
to over 1 m tall. It may otherwise be
recognized by having 6 carpels and
ovoid-oblong fruits about 5 mm long.

Triglochin maritima

Occasional. Similar places and fresh marshes; DE–Nfld. May–Aug. *T. palustris* L. is of intermediate size, has 6 perianth segments, 3 carpels, and fruits 6–9 mm long. Rare. Brackish places; NY–Greenl. May–July.

ALISMATACEAE
Water-plantain Family

Common Arrowhead; Duck-potato
Sagittaria latifolia Willd.

[319]
Seaside members of this genus may be recognized by having 3 green persistent sepals, 3 white delicate petals, unisexual flowers with male flowers above any female ones, or all male flowers, many separate stamens, many separate spirally arranged pistils on a dome-shaped receptacle. Fruits are achenes. Leaves vary from linear to those with blades broad, some with somewhat cordate to sagittate bases. Leaves of any one species may be quite variable, often on the same plant. Individual plants are often difficult to name to species.

This species may be recognized by its long-petioled leaves with hastate to sagittate lobes on base of blade. Bracts below the flowers are much shorter than the pedicels. Stalks of the fruiting heads are spreading to ascending. Plants usually much more robust than one in photograph, with inflorescences to 80 cm tall, including peduncle. Overwintering by corms developed on ends of slender rhizomes. Occasional. Pond edges, freshwater marshes, swamps, sloughs, ditches; eTX–Nfld. May–Sept. Syn.: *S. pubescens* Muhl.

Lance-leaved Sagittaria
Sagittaria lancifolia L.

[320]
A perennial with elliptic to lance-elliptic or oblong-elliptic leaf blades to 40 cm long and 10 cm wide, petioles to 80 cm long. Inflorescence to 1.5 m tall, including peduncle. Stamen fila-

Sagittaria lancifolia

Sagittaria lancifolia

ments cobwebby-hairy and linear throughout. Stalks of fruiting heads spreading to ascending. Common. Ponds, ditches, sloughs, swamps, freshwater marshes; eTX–MD. May–Nov. Syn.: *S. falcata* Pursh.

Narrow-leaved Sagittaria
Sagittaria graminea Michx.

[321]
Variable perennial, entirely submerged, partly submerged, or entirely emersed. Individual plants may exhibit all these conditions as water levels alter. Leaves linear or some emersed leaves with linear-lanceolate to elliptic blades 0.5–3 cm wide and to 50 cm long. Stamens over 2 mm long, the filaments dilated at base. Stalks of fruiting heads spreading or ascending. Common. Ponds, swamps, freshwater marshes, ditches, sloughs; eTX–Nfld. Apr.–Nov.

Three other species may be encountered in seaside habitats. They may be submerged at high water levels to exposed when the water is low. Two have recurved fruiting pedicels. They are often difficult to separate from each other and vegetatively often confused with *S. graminea*. One is *S. subulata* (L.) Buch. with linear leaves 5–30 cm long and 3–20 mm wide. Occasional. Fresh to brackish water—sloughs, tidal marshes; AL–MA. May–Sept. The other, *S. stagnorum* Small, also has linear leaves but frequently also has petioled leaves with floating elliptic to oval or ovate blades. Rare. Ponds, swamps, ditches; AL–VA. May–Sept. The third species, *S. teres* S. Wats., is recognized by having erect terete leaves to 60 cm long. Rare. Pond margins, swamps; NJ–MA. July–Sept.

XYRIDACEAE
Yellow-eyed-grass Family

Yellow-eyed-grass
Xyris iridifolia Chapm.

[322]
Xyris species are easy to recognize. Flowers are in axils of woody scales, arranged in a compact terminal spike on a leafless stem. Petals yellow, 3. Leaves linear to terete-filiform, the bases gradually to abruptly widened and overlapping, attached so that all leaves are in one plane, as in *Iris* species. Flowers are perfect, regular, 0–3 opening daily from each spike over a considerable period of time. Stamens 3. Ovary superior. We have recorded 15 species in seaside habitats. Most are rare. One or more occur from eTX–sME. Identification to species is often quite difficult, frequently being much dependent on minute characters of the seeds and the peculiar sepals.

This species is fairly easy to recognize. Leaves are linear, 40–70 cm long, 10–25 mm wide, bases keeled and pinkish to purplish. Stem flattened and 2-edged. Fruiting spike usually over 15 mm long. Flowers opening in the morning. Sepals shorter than the scales. Occasional. Wet compact soils—roadsides, pond margins, low thin pinelands, depressions; MS–wFL; GA–sNC. July–Sept.

X. caroliniana Walt. is also fairly easy to recognize. Leaf bases chestnut-brown, fruiting spikes narrowly ellipsoid to lance-ovoid, 15–30 mm long, sepals protruding beyond ends of bracts, flowers opening in afternoon. Occasional. Wet to dry places in thin woods or in the open; MS–NY. June–July.

ERIOCAULACEAE
Pipewort Family

Pipewort; Buttonrods; Hatpins
Eriocaulon decangulare L.

[323, 323a]
Members of this genus are conspicuous when in flower because of the dense head of tiny flowers ("buttons") on the tip of leafless stems. All species are perennials with basal linear leaves that possess air spaces visible to the naked eye. Roots are jointed and essentially unbranched. Each of the 2 petals has a small jet-black gland near the tip.

In this species the stem is finely 8–12 ridged, to 1.1 m tall. The leaves are longer than the sheath around base of the stem. Heads hard, mature ones 10–20 mm broad. Occasional. Wet sandy or peaty soils—ponds, ditches, thin pinelands, boggy areas; eTX–sNJ. June–Oct.

E. compressum Lam. is similar but the heads are soft, easily compressed when squeezed between fingers, and the leaves are shorter than the sheath around base of the stem. Occasional. Similar places; MS–sNC. June–Oct. *E. septangulare* With. has the same aspect as the above two species but is smaller, to 25 cm tall, when in water to 1 m tall. Mature heads are only 4–5 mm broad. Occasional. Pond margins, ditches, bogs; MD–Nfld. July–Sept.

BROMELIACEAE
Pineapple Family

Spanish-moss
Tillandsia usneoides (L.) L.

[324]
This self-sustaining relative of the pineapple is an epiphyte on trees and sometimes hangs from other objects of support such as telephone lines and fences. Plants superficially resemble some mosses and lichens but Spanish-moss is a flowering plant.

Stems are slender and wiry, the leaves filiform. Both bear numerous small silvery-gray scales that are important in trapping water and nutrient-providing dust particles necessary for life. A flower and young fruit are shown in the picture. Plants have been used for forage and a filler in upholstery and mattresses. Campers should be wary of using spanish-moss as bedding because it harbors redbugs (chiggers). Common. Pendant on trees and sometimes other objects from swamp to upland; TX–sVA. Apr.–July.

Air-plants
Tillandsia setacea Sw.

[325]
These plants are epiphytes on trees. Leaves very narrow, flat and tapering gradually to a filiform tip, dilated at base, the margins usually inrolled, the leaf appearing terete at a casual glance. Leaves in compact basal tufts, generally straight or little curved, often longer than the flowering and fruiting stalks, bearing numerous small silvery-gray scales. Flowers several in terminal dense spikes, the peduncles over 1 mm wide and bearing conspicuous bracts. Petals erect, violet, about 25 mm long. Rare. Mostly on live oaks, maritime oaks, and southern magnolia; FL–GA. June–Sept. Syn.: *T. tenuifolia* L.

T. recurvata (L.) L. is similar but there are only 1 or 2, rarely to 5, flowers at end of flowering stem, the leaves are curved and not dilated at base. Peduncle without bracts except at base, wiry, under 0.5 mm wide. Rare. Similar places; eFL–sGA. June–Sept.

COMMELINACEAE
Spiderwort Family

Dayflower
Commelina erecta L.

[326]
Seaside members of this family have 3 separate ephemeral petals that sometimes wither before noon. Sepals green, 3, separate. Stamens 6, all fertile, or fertile and sterile mixed. Leaves linear to ovate, alternate, with tubular sheaths at base. Fruit a capsule with 1–2 seeds per carpel.

This species is a perennial from thickened roots. Two petals blue, the third whitish and smaller than the other 2. Flowers appear singly each day or so from within the fold of greenish bracts called spathes. These are terminal or axillary and the margins not fused. Common. Dry usually sandy soils, usually in the open—around buildings, thin woods, roadsides; eTX–sVA. May–Oct. Syn.: *C. angustifolia* Michx.

C. communis L. also has 1 pale petal. It is a weedy annual with fibrous roots. The spathes are fused at the base. Common. Moist places in thin woods or in the open—ditches; depressions in gardens, yards, old fields; eTX–MA. May–Oct. *C. diffusa* Burm. f. has a similar aspect but has all 3 petals blue. The spathes are not fused. Stems are much-branched, decumbent, and rooting at the nodes. Common. Moist to dry places in woods or in the open; TX–sVA. Apr.–Nov.

Spiderwort
Tradescantia virginiana L.

[327]
Erect somewhat succulent perennial to 60 cm tall, not glaucous. Leaves linear, blades of the upper leaves narrower than to about as broad as the basal sheath, which is unfused. Pedicels and sepals are hairy their entire length, the hairs glandless. Petals are blue to purple or purplish-pink, or rarely white. Stamens 5 or 6, the filaments prominently bearded. Fruit a 3- to 6-seeded capsule. Occasional. Usually well-drained areas—stable dunes, roadsides, thin woods, around buildings; MS–AL; NJ–sME. Mar.–July.

T. ohiensis Raf. is easily recognized. Stem and leaves are glaucous, and pedicels and sepals, except perhaps the tips, are not hairy. Common. Stable dune areas, roadsides, fencerows, thin woods, waste places; eTX–sVA. Mar.–July.

PONTEDERIACEAE
Pickerel-weed Family

Water-hyacinth
Eichhornia crassipes (Mart.) Solms

[328]
Often seen floating on water in such a dense mass of deep-green foliage that the water and the distinctive inflated petioles are not easily seen. Largest leaves on any one plant may be only a few centimeters long, or reach a meter. Flowers are in a dense erect spike. The plant masses are a serious problem in drainage ditches, canals, ponds, etc. Its abundance diminishes greatly in the cooler parts of its range. Occasional. On fresh or mildly brackish water, or mud or muck during periods of low water; TX–sSC. Apr.–Sept.

Pickerel-weed
Pontederia cordata L.

[329]
A soft-stemmed perennial to 1 m tall. One leaf not far below the flowers, the others basal. Leaves cordate to lanceolate or rarely linear, the largest on any one plant 1–14 cm wide. Perianth purplish-blue or rarely white, the upper segment with a yellow area. The roots produce a severe burning sensation on the tongue when eaten. Common. Usually in the open—at margin of, or in, water; especially in ditches, freshwater marshes, ponds; eTX–NS. Mar.–Oct.

P. lanceolata Nutt., with narrower leaves, has been separated as a species and a variety, but intergradation of characters is too great for taxonomic recognition at either level.

JUNCACEAE
Rush Family

Needle Rush
Juncus roemerianus Scheele

[330]
Rushes are important constituents of seaside habitats in terms of both numbers of individuals and species. They are grasslike or sedgelike perennials (one species is an annual) but with a regular chaffy or scalelike perianth that is green to brown, reddish, or almost reddish black. Sepals and petals are essentially alike but in 2 separate whorls. Stamens 6, or less commonly 3. Ovary superior. Fruit a 3-carpelled capsule. Leaves have sheaths that are split longitudinally. Blades are terete to flattened, and filiform to linear, or sometimes absent. Inflorescences are terminal on simple stems and are subtended by a bract that may be small and inconspicuous to prominently extended so that the bract appears like a continuation of the stem, the inflorescence thus appearing lateral. Flowers may be a few to abundant, arranged singly or in clusters of 2 to several. Clusters may be few to many. Individual flowers are difficult to recognize in some clusters. Identification to species is often dependent on minute details and is sometimes difficult even with aid of a microscope and detailed illustrations. We have recorded 35 species in seaside habitats. We will describe a few of the more important ones and give some idea of the variation in the genus.

Needle Rush is one of the most important plants in salt and brackish marshes and in these habitats is easily recognized by the lateral-appearing inflorescence and hard, sharp point on the terminal bract. These points easily puncture the skin, a common occurrence for persons walking among the plants. Common. Upper portions of salt and brackish marshes, often in solid stands; eTX–MD. Mar.–Oct.

Juncus roemerianus

Soft Rush
Juncus effusus L.

[331]
Stems usually in large tussocks, developing from a single stem. Stems with fine straight longitudinal lines traceable for several cm. Leaf sheaths lacking blades. Inflorescence lateral-appearing. Flowers borne singly, not in clusters as in the preceding species. Perianth segments all sharp-pointed, equaling the capsule. Common. Freshwater habitats—marshes, ditches, pond margins, depressions in open; eTX–Nfld. July–Sept.

J. balticus Willd. is another species, with lateral-appearing inflorescences that may be encountered in seaside habitats. The stems lack longitudinal lines or, if present, the lines are indistinct and irregular. Occasional. Freshwater and brackish marshes, swales, sloughs; NY–Nfld. May–Sept.

Juncus effusus

Black-grass
Juncus gerardi Loisel.

[332]
This is one of the many species in which the inflorescence is obviously terminal. Leaf sheaths mostly with blades. Some leaves arising within the upper ¾ of the stem. Flowers arranged singly. Sepals 2.4–3.2 mm long. Anthers about 3 times the length of filaments. Capsule about as

long as sepals. Spreading by rhizomes and slender stolons, forming extensive mats. Common. Salt and brackish marshes and shores; VA–Nfld. June–Sept.

J. bufonius L. is another species with an obviously terminal inflorescence that is fairly easy to identify. It is unusual in being an annual. Flowers are borne singly as in the above species but the inflorescence is ⅓ or more of the plant; plants to 35 cm tall, and sepals 3.5–6.6 mm long. Common. Moist places in the open—roadsides, around buildings, freshwater marshes; TX–Nfld. June–Nov.

Juncus gerardi   *Juncus bufonius*

Path Rush
Juncus tenuis Willd.

[333]
Plants 10–60 cm tall, leaves all basal, blades flat, sometimes rolled and seemingly terete when dry. Leaf sheath margins thin, whitish-transparent, friable. Flowers borne singly. Capsule with rounded summit. Common. Margins of paths, vehicle trails, ponds, and swamps; depressions; eTX–Nfld. June–Sept.

J. dichotomus Ell. is similar, to 80 cm tall, the flowers borne singly, but the margins of the leaf sheaths are firm and not transparent, the blades subterete and grooved, and the bract at base of inflorescence shorter than in

Juncus tenuis

Path Rush. Common. Marsh margins, depressions in thin woods, ditches, swales; eTX–MA. June–Oct.

Large-headed Rush
Juncus megacephalus M. A. Curtis

[334]
Over half of the rushes in seaside habitats have flowers in a few to many compact heads such as this species. Individual flowers are usually difficult to recognize and identification to species is usually troublesome, to say the least.

This rush grows to 1.2 m tall, spreading by basal offshoots. Blades of the midstem leaves are terete with cross partitions, the uppermost leaf with a blade much shorter than the sheath, lower leaf sheaths bladeless. Capsule gradually tapering towards the summit, a little longer than the perianth. Seeds are needed for positive identification. They are elliptical and lack long tail-like extremities. Common. Moist to wet places, often in shallow water—swales, sloughs, ditches, pond margins, edges of freshwater or brackish marshes; eTX–MD. June–Aug.

J. scirpoides Lam. is similar in many ways but spreads by short stout rhizomes and the blade of the upper stem leaf equal to or longer than its sheath. Common. Similar places and in wet grasslands; eTX–NY. June–Oct.

Juncus megacephalus

Flat-leaved Rush
Juncus validus Cov.

[335]
Plants to 1 m tall. Blades of lower leaves flattened and with cross partitions. Capsule tapering toward the tip, slightly longer than perianth, the sections separating at tips when splitting open. Seeds broadly elliptic, lacking long tail-like extremities. Common. Freshwater or brackish marshes, ditches, pond margins, depressions in the open; eTX–sNC. July–Sept.

J. polycephalus Michx. is similar but a little more robust, and the sections of the capsule remain united at their tips when splitting open. Occasional. Similar places; eTX–sNC. July–Sept. *J. acuminatus* Michx. has characteristics similar to those of the above 2 species, except the capsules are abruptly pointed and about as long as the perianth. Common. Similar places; eTX–sME. May–Aug.

J. marginatus Rostk. also has flowers in compact heads but the heads are smaller, usually more numerous, and closer together; the leaf blades are flat and lack cross partitions. Common. Similar places; TX–NS. June–Sept. Syn.: *J. biflorus* Ell.

Juncus marginatus

Juncus validus

LILIACEAE
Lily Family

Wild Onion; Canada-garlic
Allium canadense L.

[336]
Onions are easily recognized by their strong onion odor when the leaves are bruised, bulbous bases, and flowers in umbels. Fruits are 3-carpelled capsules.

This onion usually has bulblets replacing some or all flowers. Leaves are mostly basal, smooth and flat, 2–6 mm wide. Bulbs are coated with netted fibers. Peduncles are straight. Perianth segments acute, 5–7 mm long. Plants to 55 cm tall. Occasional. Roadsides, fields, waste places, around buildings, yards; eTX–sME. Mar.–July.

Cylindrical glaucous hollow leaves identify *A. vineale* L. (Field-garlic). Plants to 80 cm tall. Occasional. Similar places; NC–MA. May–July.

False-garlic
Nothoscordum bivalve (L.) Britt.

[337]
Perennial to 45 cm tall from a small bulb that has a faint odor of onion when fresh. Leaves linear, flat, less than 5 mm wide. Flowers 3–10 in a terminal umbel. Perianth segments 9–15 mm long, reflexed when fruits are developed. Onion bulbs have a strong odor and the perianth is not reflexed. Common. Thin woods, roadsides, lawns, swales, grasslands, stable dune areas; TX–VA. Mar.–May; Sept.–Nov.

Wood Lily
Lilium philadelphicum L.

[338]
Lilies are uncommon in seaside habitats but if encountered are unforgettable. Wood Lily is the species most likely to be encountered. Plants to 1 m tall, leaves whorled, flower erect, ovary superior, fruit a 3-carpelled capsule. Rare. Thin woods, meadows, low shrub areas; NY–ME. June–Aug.

There is one seaside record for the equally beautiful *L. catesbaei* Walt. (Pine Lily). It is similar to Wood Lily but has alternate leaves. Rare. Open wet habitats; most likely from MS to FL. July–Sept.

Starry False-Solomon's-seal
Smilacina stellata (L.) Desf.

[339]
Erect to arching perennial to 65 cm tall. Flowers in racemes. Sepals and petals alike, white or nearly so, all separate, 4–6 mm long. Maturing fruits are green berries with dark maroon stripes, later becoming nearly uniformly dark red; 6–10 mm thick. Occasional. Stable dunes, thickets, thin woods; NJ–Nfld. May–Aug.

Canada Mayflower; Wild Lily-of-the-Valley
Maianthemum canadense Desf.

[340]
Erect perennial to 25 cm tall. Leaves usually 2, occasionally 3, rarely 1, on flowering plants. Flowers in a small raceme. Perianth segments 4, separate, alike, about 2 mm long, reflexed. Mature fruits are reddish berries 3–6 mm in diameter. Common. Woods, thickets, bogs, between mature dunes; NJ–Nfld. May–June.

AMARYLLIDACEAE
Amaryllis Family

Yellow Star-grass
Hypoxis hirsuta (L.) Cov.

[341]
Perennial from a corm surrounded by a few thin pale to brownish sheaths. Leaves basal, hairy, grasslike, 3–12 mm wide. Flowering stem to 35 cm tall. Flowers 2–9. Sepals obtuse. Ovary inferior. Fruit a capsule. Seeds black. Occasional. Thin pinelands, dry grasslands, roadsides; eTX–sME. Mar.–Sept.

H. micrantha Pollard is similar but the basal sheath is fibrous, flowers only 1–2, sepals acute, and seeds brown. Rare. Thin pinelands, grasslands; AL–sNC. Mar.–Apr.

IRIDACEAE
Iris Family

Blue-flag
Iris versicolor L.

[342]
Perennial to 1 m tall from a thick rhi-

zome, often forming large clumps. Leaves 5–30 mm broad and, as in all irises, flattened into one plane, at least at their bases. Petal-like structures of the flower are 9. The broadest and lowest 3 are the sepals, which are entirely glabrous and are 20–35 mm broad; the next three, the nearly erect structures in the photograph, are the petals, which are 5–20 mm wide and shorter; the other three, which are narrowest, are the stigmas and may be seen projecting between the sepals and petals in the photograph. The 3 stamens are hidden under the stigmas. The ovary is inferior, 3-carpelled. Ovary and capsule bluntly angled. Occasional. Freshwater or brackish marshes, ditches, wet meadows; nNJ–Nfld. May–July.

I. virginica L. is similar but the sepals have short hairs on the upper surface at the base. Common. In wet places or shallow water, freshwater marshes, ditches, sloughs, ponds, depressions in thin pinelands or broadleaf woods; eTX–sVA. Apr.–May. *I. prismatica* Pursh is easily recognized by its sharply angled ovary and fruit. Leaves are only 3–7 mm wide. Occasional. Freshwater and brackish marshes, wet meadows, depressions; DE–NS. May–June.

Blue-eyed-grass; Sisyrinchium
Sisyrinchium albidum Raf.

[343]
Sisyrinchiums are easily recognized by their stems and leaves that are flattened and similar in size and in dense clumps; sets of leaves aligned into one plane, at least at base; 6 separate similar perianth segments; 3 stamens united by their filaments; and inferior ovaries. Fruits are 3-carpelled capsules. Identification to species is sometimes difficult. We have recorded 8 species in seaside habitats, 2 annuals and 6 perennials. Most individuals of annuals are under 20 cm tall, whereas most perennials are over 20 cm tall.

This species is the only seaside perennial having a pair of bracts under the flower cluster and the only one with the edges of the outer of the 2 bracts free to the base. Plants to 40 cm tall, perianth whitish to pale violet. Common. Usually dry places— thin pinelands, pine-palmetto woods, grasslands, roadsides, stable dune areas; eTX–GA. Mar.–May.

S. arenicola Bickn. is similar and easily recognized by old leaf bases persisting as tufts of crowded bristlelike fibers. Bracts at top of stem are 2–5 and not paired. Rare. Similar places; MS–NS. Apr.–July.

Sisyrinchium albidum

Blue-eyed-grass
Sisyrinchium atlanticum Bickn.

[344]
Perennial to 50 cm tall. Stems 1–3 mm wide, one of few species with stems that narrow; 2 or more stalked flower clusters at top of principal stem, outer bract under flower clusters with edges fused near their bases; capsule 2–4 mm long. Common. Moist to wet areas in the open—ditches, freshwater marshes, meadows, pond shores; eTX–NS. Mar.–June.

S. mucronatum Michx., another perennial, also has narrow stems, 0.5–1.5 mm wide. Plants to 40 cm tall. May be distinguished from the above species by having only 1 flower cluster at top of each stem. Occasional. Moist places—meadows, thin woods, between stable dunes, pond shores; nGA–sME. May–June.

The 2 other seaside perennials, which are similar, have wider stems, 3–5 mm, and longer capsules, 4–6 mm. In *S. angustifolium* Mill. mature fruits and dried plants are pale-colored. Common. Similar places; nNC–Nfld. Mar.–July. In the other species, *S. montanum* Greene, mature fruits are dark and dried plants are blackish. Occasional. Similar places; nNJ–NS. June–July.

Sisyrinchium atlanticum

Annual Blue-eyed-grass
Sisyrinchium exile Bickn.

[345]
An annual to 20 cm tall. Perianth yellow or light cream-colored with a brownish red eye ring. Occasional. Usually in moist to wet places—roadsides, lawns, meadows, old fields; eTX–wFL. Mar.–May.

S. *rosulatum* Bickn. is similar but its perianth varies from white to lavender-rose and has a rose-purple eye ring. Common. Similar places; eTX–sNC. Apr.–May.

CANNACEAE
Canna Family

Golden Canna
Canna flaccida Salisb.

[346]
Perennial to 1.3 m tall from coarse rhizomes. Leaves lanceolate-ovate, tapered to sheathing bases, to 55 cm long and 15 cm wide, lower leaves often petioled. Sepals 3, greenish, 25–30 mm long. Petals 3, yellowish green, their bases united into a tube 50–65 mm long, the lobes about the same length. The showy part of the flower consists of modified stamens that look like petals. One petaloid stamen bears the only pollen-bearing anther. The style is single and also petaloid. The inferior ovary and fruit are covered with small elongated warts. Rare. Freshwater swamps and

marshes, pond edges; TX–sSC. Apr.–Aug.

ORCHIDACEAE
Orchid Family

Pink Lady's-slipper; Moccasin-flower
Cypripedium acaule Ait.

[347]
In orchids, 2 petals are similar, the other (the lip) is different, often radically so, and is an important feature in identification. Another peculiarity of orchids is the union of the 1–3 stamens and the pistil, forming much of the column, the central structure of the flower. All have inferior ovaries. Fruits are 3-carpelled capsules containing enormous numbers of minute seeds.

C. *acaule* gets its common name from the inflated moccasinlike lip. It is a hairy perennial to 55 cm tall, with 2 basal leaves. Occasional. Dry to wet places—often closely associated with roots of pine, stable wooded dunes, pine deciduous woods, bogs, swamps; VA–Nfld. Apr.–June.

Spring Ladies'-tresses
Spiranthes vernalis Engelm. & Gray

[348]
A variable genus, with plants often difficult to name to species, diagnostic characters often being small. The name *Spiranthes* means "coil-flower" in allusion to the spiral arrangement of the flowers. We have recorded 5 species in seaside habitats.

This species is a perennial to 1.1 m tall from several thick roots. Upper parts with dense fine-pointed glandless hairs. Leaves few, basal, erect to ascending, up to 30 cm long and 1 cm wide, their bases sheathing the stem. Flowers strongly spiraled, rarely as few as twice around the stem. Lip, the lowest perianth segment, 4.5–8 mm long, widest near base. Common. Moist places—roadsides, thin pinelands, freshwater and brackish marshes, swales, low-energy beaches, prairies; eTX–MD. Mar.–July.

S. praecox (Walt.) Wats. has a similar aspect but upper parts of the plant are glabrous or with sparse bulbous-tipped hairs, and the lip is oblong to elliptic, sometimes widest at the tip, usually finely veined with green. Common. Moist to wet places—freshwater marshes, wet grasslands, thin pinelands, swales, meadows; eTX–NY. Mar.–July.

Autumn Ladies'-tresses
Spiranthes cernua (L.) Rich.

[349]
Plants glabrous below, finely downy-hairy above, flowers usually fragrant, in several spirals forming a dense spike 2–3 cm wide. Lip 6–14 mm long and with 2 small rounded projections at its base. The largest leaves are near the base of the stem, linear to lanceolate, 5–40 cm long, 5–20 mm wide. Common. Moist to wet places—thin pinelands, swales, swamps, thin scrub, roadsides; eTX–NS. July–frost.

S. tuberosa Raf. is fairly easy to recognize. It has petioled and spreading basal leaves, a single tuberlike root, the lip under 4 mm long, the spike extremely slender, and flowers varying from a single row to strongly spiraled. Occasional. Broadleaf woods, old fields, thin shrub areas; eTX–MA. June–Oct. Syn.: *S. grayi* Ames. *S. gracilis* (Bigel.) Beck also has spreading petioled leaves but the lip is 4–6 mm long and there is usually more than one tuberous root. Rare. Broadleaf woods, old fields, meadows, low pinelands; eTX–wFL; nNC–sVA; NH–NS. July–Oct.

Coral-root
Corallorhiza wisteriana Conrad

[350]
These nongreen orchids obtain their food from dead plant remains, probably with the aid of fungi. Plants to 40 cm tall, inconspicuous, varying from tan (usually only at base) to dark reddish purple and blending with the leaves on the ground. The corolla lip is white, pendant, 5–7 mm long, and conspicuously spotted with magenta-purple. Occasional. Light to rich, dry to moist soils in various kinds of woods; eTX–GA. Feb.–Apr.

Tree Orchid; Green-fly Orchid
Epidendrum conopseum R. Br.

[351]
Probably more abundant than known records indicate, as plants, which are attached to tree limbs, often escape notice, especially those camouflaged by resurrection fern and spanish-moss. Plants are epiphytes using the tree mainly for support, perhaps obtaining some nutrients from the bark. Plants to 40 cm tall, stems slender. Roots thick, spongy, matted. Leaves 1–3, leathery, linear-lanceolate to narrowly oblong, sessile. Flowers fragrant, in racemes, grayish-green, occasionally tinged with purple. Rare. Mostly on live and maritime oaks, and the southern magnolia; LA–sSC; can be expected in TX and into extreme sNC. July–Sept.

Herbaceous Monocots without Sepals and Petals

TYPHACEAE
Cattail Family

Common Cattail
Typha latifolia L.

[352]
Erect coarsely rhizomatous perennial to 3 m tall. Leaves 2-ranked, glabrous, long-linear, nearly flat, with cylindrical basal sheaths that taper into the blade. Flowers tiny, lacking a perianth, unisexual, numerous, in dense spikes, the mass of male flowers above that of the female flowers, usually without a space between the male and female areas. Mature fruit-

ing spikes reddish brown, 10–18 cm long, about 6 times as long as thick. Rhizomes, young stems, flowers, and pollen were important sources of food for the Indians. Common. Shallow freshwater or wet places—marshes, ponds, ditches, swales, sloughs; TX–Nfld. Apr.–July.

T. angustifolia L. is similar but leaves are rounded on the back, sheaths usually auricled at junction with blade, staminate and pistillate masses of spike usually separated by an interval, and mature fruiting spikes 8–20 cm long, slender, 6–10 times as long as thick. Common. Brackish as well as freshwater habitats; TX–NS. Apr.–July. *T. domingensis* Pers. (Tule) is much like the latter species but mature spikes are 15–25 cm long, about 10 times as long as thick, and leaf sheaths tapering into blade. Common. Brackish and freshwater habitats; TX–DE. May–July. The species hybridize.

POACEAE
Grass Family

Grasses are a substantial, highly visible, valuable component of practically all seaside habitats, and a study that excludes them is incomplete. They do have a reputation, often justified, for being difficult to identify, and some members of other families are occasionally confused with them. Many seaside species, however, are easily recognized, as a little attention to them will demonstrate, and identification of the ones included in this book is simplified because of the limited number of species in the habitats involved.

Grasses have leaves that are 2-ranked, a condition usually quite evident. In seaside grasses, the leaf consists of 2 major parts: the basal portion, called a sheath, that closely surrounds the stem and has margins with free edges that often overlap or, rarely, are fused; and the upper portion, the blade, which diverges from the stem and is mostly long-linear.

Grass flowers lack a perianth and, except for one microscopic part that is unimportant here, consist of 3 (rarely 1 to 6) stamens, or 1 pistil, or both stamen(s) and pistil. These are inserted between and at the base of 2 "bracts"; the outer and the larger "bract" is known as the lemma, the inner and often inconspicuous one is known as the palea. The lemma, palea, stamen(s), and/or pistil combined are known as the floret.

Florets are sessile and attached singly or in alternate 2-ranked sets of 2 to many. In most grasses at the base of the single floret or of the 2 to many florets are 2 empty (thus sterile) "bracts" called glumes. These together with the floret(s) constitute the spikelet. Spikelets may be sessile or pedicellate and are usually arranged in panicles. Some are pedicellate and in racemes or spikelike racemes. Other species have sessile spikelets that are arranged in spikes, some 2-sided, some 1-sided.

The fruit of a grass is a grain, or in two seaside genera (*Sporobolus, Eleusine*) it has a thin wall with the enclosed seed fastened only by its stalk.

In identifying grasses, much help is obtained by noting certain easily recognized characteristics, such as spikelet size; whether or not the spikelets have pedicels or are sessile; the number of florets per spikelet; the kind, size, shape, and compactness of the inflorescence bearing the spikelets; and the presence, location, and character of hairs. Some grasses have a hard, shiny, smooth lemma; a few species have sharp spines associated with the spikelets; others have bristles mixed with spikelets; and many have awns on lemma tips. A 10 × hand lens is often useful.

Seaside species that are vegetatively similar to grasses are predominantly members of the Juncaceae (Rushes), Liliaceae (Lilies), Amaryllidaceae (Amaryllises), Commelinaceae (Spiderworts), Iridaceae (Irises), Xyridaceae (Xyrises), Asteraceae (Composites), Cyperaceae (Sedges), Typhaceae (Cattails), and Eriocaulaceae (Pipeworts).

In the first 7, possibly confusing, families, the perianth is sufficiently developed to be recognized, usually readily, and fruits are other than grains.

Members of the last 3 families are easily separated from grasses by features specific to the individual families. Pipeworts have a perianth that is difficult to recognize, but the tiny flowers are in conspicuous dense heads on the tip of a leafless stem. The tiny flowers of cattails are arranged at the top of the stem in huge numbers in coarse cylindrical clusters. Sedges are the most similar to grasses and can be separated by the characters given below. An asterisk (*) indicates there are a few exceptions.

POACEAE (see p. 61 for key)	CYPERACEAE (see p. 61 for key)
1. Leaves 2-ranked	1. Leaves 3-ranked
2. Leaf sheaths with free margins*	2. Leaf sheaths with margins fused
3. Internodes hollow*	3. Internodes solid*
4. Stem cross section circular except for auxillary bud notch	4. Stem cross section triangular*
5. Flower between 2 "bracts"*	5. Flower in axil of 1 "bract"
6. Stigmas 2	6. Stigmas 2 or 3
7. Fruit a grain*	7. Fruit an achene or nutlet

In using leaf rank as a distinguishing feature, attention should be focused on the leaf bases, noting whether they are in 2 rows up and down the stem or in 3 rows. Seedlings with 3 or more leaves are easily separated in this manner. This is an especially useful character to determine which of the 2 families is involved in lawn and garden weeds.

As an aid in the recognition of grass spikelets and their variations, it is helpful to call attention to the drawings accompanying some of the species illustrated, listing some specific features:
Distichlis spicata (p. 338): spikelet with 2 glumes, 9 lemmas.
Agrostis stolonifera (p. 334): spikelet with 2 glumes, a lemma.
Panicum rhizomatum (p. 329): spikelet with 2 veined glumes, a veined lemma, a firm smooth veinless lemma.
Panicum dichotomiflorum (p. 329): the same, but veinless lemma hidden.
Paspalum floridanum (p. 327): a glume, a veined lemma, the smooth veinless lemma hidden.
Elymus virginicus (p. 339): spikelet with 2 awned glumes, 3 awned lemmas.
Erianthus coarctatus (p. 324): a portion of the axis of the raceme of spikelets, a sessile spikelet, a stalked spikelet, each with an awn at tip and hairs at base.
Sorghastrum secundum (p. 326): awned spikelet, the hairy stalk of a spikelet that never develops, a portion of the axis of a raceme.
The palea, a structure often similar to lemmas, is not readily visible in the species listed above. It is often visible in others, as in the drawings of *Ammophila breviligulata* (p. 334), *Spartina cynosuroides* (p. 335), and *Sporobolus virginicus* (p. 333), which all have spikelets with 2 glumes, a lemma, and a palea evident.
Spikelets with a visible palea are easily misinterpreted as having an extra lemma. This problem can be avoided by checking for stamens and pistils, or fruits. These are situated between and fastened at the base of the lemma and palea. If there is 1 place where these occur, there is only 1 lemma, as is the case in the last 3 species above; if there are 2 places, there are 2 lemmas; 3 places, 3 lemmas.

Gamma Grass
Tripsacum dactyloides (L.) L.

[353]
Perennial from coarse rhizomes.
Plants to 2.5 m tall. Leaves largely
basal and on lower half of stem,
blades to 60 cm long and 3 cm wide.
Flowers in spikes, unisexual, the
sexes borne in different parts of the
same spike, with the male above the
female. Spikes 1 to a few, terminal on
the principal stem axis and on the few
erect lateral branches. Male spike almost identical in appearance to sections of the "tassel" of corn plants,
dropping off as a whole shortly after
shedding pollen. Individual female
flowers and grains developing from
them are in bony beadlike joints of
the spike, which breaks apart at maturity. Plants in thin clusters or colonies, or occasionally solitary. Occasional. Usually moist places—ditches,
depressions, swales, thin woods,
waste places; TX–MA. May–Nov.

Giant Sugarcane Plumegrass
Erianthus giganteus (Walt.) Muhl.

[354]
Among the tallest grasses, growing to
4 m tall, often forming clumps or
large colonies. Leaf blades to 50 cm
long and 23 mm wide. Spikelets in
dense panicles, the branches of which
are racemes with similarly paired
spikelets, 1 sessile and 1 pedicelled.
Few seaside grass genera have spikelets in pairs. Spikelets apparently
with 1 floret that bears a thin awn 20–
25 mm long and many long hairs,
those hairs at the base longer than
those near the top, and much longer
than the spikelet. At maturity the
pedicellate spikelet falls off; sections
of the raceme separate, each falling
with a sessile spikelet and pedicel of
the other spikelet. Common.
Marshes, thin woods of swamps,
swales, sloughs, pond margins,
ditches, depressions; TX–NJ. Aug.–
Oct.
    E. coarctatus Fern. is similar but
smaller, to 2.5 m tall, the panicle slimmer and not as dense, and the hairs

on the spikelet shorter than the spike-
let. Occasional. Similar places; LA–
DE. Sept.–Oct. Syn.: *E. brevibarbis*
Michx. as used in manuals.

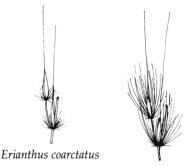

Erianthus coarctatus

Erianthus giganteus

Bushy Broomsedge
Andropogon glomeratus (Walt.) B.S.P.

[355]
Andropogon species are prominent features of seaside habitats in the fall.
They are perennials and much alike
vegetatively; differences between species mostly involve inflorescences. All
have conspicuous soft silky hairs in
the spikelet-bearing racemes. Spikelets are in pairs in racemes (unusual
arrangement in grasses); one spikelet
is sessile, the other is stalked and different in size or shape or may be absent, in which case being represented
by its pedicel. The soft silky hairs,
paired spikelets, and 2 to several racemes at end of each peduncle serve
to identify seaside *Andropogon* species. Identification to species is sometimes difficult, partly because botanists differ in their interpretation of
the species. Our treatment here
therefore differs from that in some
manuals. *Andropogon* closely resembles *Schizachyrium*, the next genus, which see for separation.
    Bushy Broomsedge grows to 1.5 m
tall. It is easily recognized by the
large dense mass of hairy inflorescences making the plant bushy-
topped. The paired racemes are short-
peduncled and set between 2 clasping
bractlike leaves. Common. Moist to
wet places in the open—swales,

Andropogon glomeratus

ditches, sloughs, meadows, freshwater marshes, margins of brackish marshes, depressions; eTX–sMA. Aug.–Oct. Syn.: *A. virginicus* L. var. *abbreviatus* (Hack.) Fern. & Grisc.

Chalky Broomsedge
Andropogon capillipes Nash

[356]
Plants to 1 m tall, usually in clumps; glaucous, especially toward the base. Inflorescences nearly uniformly distributed in the upper half or more of the plant. Racemes are shorter than the bractlike leaves clasping them. Stem branches just below these bracts are glabrous. Occasional. In thin woods or in the open—swales, stable dunes, ditches, margins of freshwater marshes, meadows, roadsides, depressions; LA–sVA. Aug.–Oct.

A. glaucopsis (Ell.) Nash is also glaucous, usually heavily so, and is otherwise much like *A. capillipes* except the stem branches just below the bracts clasping the racemes are long hairy. Occasional. Similar places; TX–sVA. Aug.–Oct. *A. virginicus* L. (Virginia Broomsedge) has a similar aspect to the above 2 species and is otherwise similar but is not glaucous. Hairs 5–7 mm long on the raceme axis and on the peduncle supporting the 2 racemes. Common. In thin woods or in the open—dunes, swales, meadows, marsh margins, roadsides, old fields, pinelands; eTX–MA. Aug.–Oct.

Sand Broomsedge
Andropogon longiberbis Hack.

[357]
Resembling *A capillipes,* but not glaucous, and *A. virginicus*. Racemes are relatively conspicuous because they are 2–3 on each peduncle and about 3 cm long. Also, hairs on the pedicels and axis of the racemes are relatively long, 7–10 mm. Occasional. Dunes, dry meadows, sandy pinelands; MS–NC. Sept.–Oct.

A. elliottii Chapm. (Elliott's Broomsedge) is an easily recognized species. It has dense clusters of racemes, but in much smaller masses than *A. glomeratus*. The raceme pairs are clasped by leaflike bracts as in several other species, but, in addition, groups of these clasped raceme pairs are in turn clasped by much enlarged overlapping leaf sheaths just below. These sheaths are 5–10 mm wide. Occasional. Old fields, thin live oak and maritime woods, pinelands, between stable dunes; eTX–VA. Aug.–Oct.

Andropogon elliottii

Splitbeard
Andropogon ternarius Michx.

[358]
This species grows to 1.2 m tall and is vegetatively much like *A. capillipes* and *A. longiberbis*. The inflorescence is much different, the raceme pairs being on peduncles 5–12 cm long with only the base of the peduncles being tightly clasped by the bractlike leaves. These "bracts" are inconspicuous because of their small size and clasping nature. The photograph is of the inflorescence of a plant just before the racemes break apart. The racemes are copiously hairy, the hairs to 9 mm long. The sessile spikelets are 7 mm long. Common. Dry places—thin woods, pinelands, stable dunes, old fields, roadsides, meadows; eFL–MD. Aug.–Oct.

Drooping Woodgrass
Sorghastrum secundum (Ell.) Nash

[359]
Perennial to 2.2 m tall with stems growing singly or more frequently in clumps. Leaves mostly basal. Inflorescence narrow 1-sided panicles to 45 cm long. Spikelets yellowish brown, shiny. Conspicuous stiff ascending hairs mostly 1–1.5 mm long are on one side of the spikelet and on a spikeletless pedicel alongside. Spikelet with an awn 25–35 mm long, twice-geniculate, flat, twisted when dry except near the tip, and easily pulled

Sorghastrum secundum

free of the spikelet when mature. Occasional. Pinelands, scrub oak-pine, live oak-pine; AL–GA. Sept.–Oct.

S. elliottii (Mohr) Nash (Elliott's Woodgrass) is much the same but spikelets are chestnut-brown and in a loose but not 1-sided panicle to 30 cm long. Occasional. Similar places and old fields; FL–VA. Sept.–Oct.

Johnson Grass
Sorghum halepense (L.) Pers.

[360]
Perennial from a whitish rhizome that is scaly at the nodes. Plants to 3 m tall. Internodes of stem solid, unlike all other wild seaside grasses. Spikelets of 2 kinds, a plump bisexual sessile one and a slim stamen-bearing pedicellate one, in short sometimes branching racemes, these in panicles to 60 cm long and 20 cm broad. Often a troublesome weed. Sometimes cultivated for forage but dangerous to grazing animals because plants develop sufficient cyanogenic compounds under certain conditions to cause prussic acid poisoning when eaten. Death has been known to occur 30 minutes after plants were eaten. Common. In almost any open habitat except in water; TX–MD. May–Oct.

Maritime Bluestem
Schizachyrium littorale Bick.

[361]
Seaside species of this genus are vegetatively like *Andropogon* species. Inflorescences and spikelets are also similar except that there is a single raceme on each peduncle.

This species is separated from others of the genus by the following: internodes of the rhizomes 5–40 mm long, stems often decumbent and usually rooting, leaves chiefly on the lower half of the plant, blades of lower leaves 2–6 mm wide. Internodes of the axis of the racemes under 1 mm across, sessile spikelet 10 mm long, pedicellate spikelet without a pistil or stamens. Occasional.

Schizachyrium littorale

Dunes, above drift lines; TX–FL; NC–NY. Aug.–Oct.

S. maritimum (Chapm.) Nash is much the same but stems are erect, more evenly leafy throughout, and the pedicellate spikelets bear stamens. Occasional. Similar places and old dunes; LA–wFL. Aug.–Oct. Syn.: *Andropogon miritmus* Chapm.

S. scoparium (Michx.) Nash (Little Bluestem) can be recognized by having internodes of rhizomes 1–3 mm long, and hairs on axis of racemes and pedicels 2–5 mm long and sparse. Common. Thin woods, old fields, well-drained meadows, stable dunes; nGA–NY. Aug.–Oct. Syn.: *Andropogon scoparius* Michx.

S. stoloniferum Nash is much like *S. littorale* but plants usually well separated because of long rhizomes, and sessile spikelet only 5–7 mm long. Rare. Habitats similar to those of *S. scoparium*; MS–GA. Sept.–Oct. Syn.: *Andropogon stoloniferer* (Nash) Hitchc.

Tall Paspalum
Paspalum floridanum Michx.

[362]
Paspalums are annuals or perennials

0.1–2 m tall. The genus is fairly easy to recognize. Spikelets are on short pedicels and are in 2 rows on a 1-sided axis; the whole is a spikelike raceme. The spikelets have one side flat, the other strongly rounded and turned toward the axis of the raceme. The inner portion of the spikelet is a hard shiny veinless light-colored floret, flat on one side and rounded on the other (called the back). The shiny floret is closely clasped by 2 similar-veined and membranaceous glumes. In reality only one is a glume, the other a sterile lemma. *Paspalum* is one of the troublesome genera; plants often are difficult to name to species. Mostly for this reason it is impractical to include here all of the 20 species we have tabulated for seaside habitats. We include and attempt to describe sufficiently for identification the most conspicuous of the species.

Tall Paspalum is a coarse erect little-branched perennial to 2 m tall with inflorescences well above the leaves. Plants can hardly be missed. Stems are leafy nearly throughout. Spikelets are about 4 mm long, only one other tall seaside paspalum having spikelets as long. Inflorescences have 2–6, rarely 7, stiff racemes 4–15 cm long. Common. Dry to wet places in thin woods or in the open—pinelands, edges of freshwater marshes and ponds, roadsides, old fields; eTX–MD. Aug.–Oct. Syn.: *P. giganteum* Bald. ex Vasey.

P. difforme Le Conte also has spikelets about 4 mm long but some are as short as 3.2 mm, plants are not as tall, and leaves are mostly at the base. Occasional. Similar places; LA–SC. Aug.–Oct.

Paspalum floridanum

Bahia Grass
Paspalum notatum Flügge

[363]
The V-shaped inflorescence high above the basal tufts of leaves makes Bahia Grass conspicuous. In addition, plants are usually in dense mats formed largely by means of coarse vigorous rhizomes. Plants grow to 80 cm tall. The inflorescence consists of the 2 racemes, each of which arises from the tip of a peduncle. Occasionally there is another raceme just below, or rarely 2 below. Spikelets are about 3.5 mm long, rarely to 4 mm. Abundantly planted for forage and along highways, commonly escaping. Common. In almost any habitat except dense woods, in water, or on the most active dunes; TX–sNC. June–Oct.

Knotgrass
Paspalum distichum L.

[364]
This species certainly is not conspicuous for its size since flowering stems are rarely over 40 cm tall. However, it frequently occurs in thick conspicuous mats, or carpets, which are formed largely through extensive rhizome growth. Knotgrass may be separated from other similar species by its loose leaf sheaths, 2 racemes at end of a leafy stem, and spikelets 2.5–4.0 mm long with inconspicuous short soft fine hairs. A closer photograph than shown is necessary to show the hairs. Occasional. Swales, accretion areas, pond margins, marsh margins, and depressions; and poorly drained spots in pathways, roads, lawns, golf fairways; TX–sVA. June–Aug. Syn.: *P. vaginatum* Sw.

Axonopus affinis Chase has a similar inflorescence and forms extensive carpets, but the spikelets are sessile, about 2 mm long, and their backs are turned away from the axis of the spikes. Spikes are slimmer and to 9 cm long. A third and sometimes fourth spike occurs below the top 2. Most leaf tips are blunt. Common. Wet or poorly drained places—live oak woods, pond margins, ditches, thin woods; occasionally in dry habitats; eFL–sVA. June–Oct.

Vasey Grass
Paspalum urvillei Steud.

[365]
Perennial to 2 m tall, erect, branching at the base and forming tufts. Inflorescence composed of 7–20 ascending to arched-spreading racemes on a single axis. Additional inflorescences are developed throughout the growing season on ends of erect branches from leaf axils. Identity is assured if spikelets are 2–3 mm long and bear conspicuous silky hairs. Plants often can be located from a distance because of movements caused by birds seeking the spikelets for food. Common. In most moist to wet places in the open, or occasionally in drier habitats; eTX–VA. May–Oct.

P. dilatatum Poir. (Dallis Grass) is easy to recognize because it is the only other seaside *Paspalum* having spikelets with long silky hairs. Spikelets are larger, 3–3.5 mm long. Stems are in clumps, often decumbent, leaves are mostly basal. Common. Very weedy and can be expected almost anywhere in the open—most frequently at roadsides, in lawns and fields, doing best in moist places; TX–VA. May–Oct.

Paspalum urvillei

Switch Grass
Panicum virgatum L.

[366]
Although spikelets of various species of *Panicum* vary in size (1–8 mm long) and in shape, the genus is easily recognized. Spikelets are in panicles and consist of (a) a hard shiny smooth fertile inner lemma that is rounded on

the back and on the other side encloses the similar-surfaced palea, and (b) 3 veined membranous units clasping the lemma, the inner 1 being a sterile lemma and the outer 2 being glumes. Identification beyond genus is another matter; only a few specialists can name most species with confidence. *Panicum*, as described above, has been split by some botanists into 2 genera, probably correctly so. The genera are *Panicum* and *Dichanthelium*, but for simplicity it is better here to consider all seaside species as being in *Panicum*. Of the 48 species we have recorded in seaside habitats, we are including here only 5; these are reasonably easy to recognize, especially when the habitats in which they occur are considered.

Switch Grass is a perennial to 2 m tall with coarse scaly rhizomes. Stems erect to arching, usually growing in clumps. Spikelets 3–5 mm long, on long pedicels in large thin panicles 15–50 cm long and nearly as wide. Common. At and in margins of marshes and ponds, depressions, thin live oak woods, swales, wet pinelands; TX–sNS. June–Oct.

Seaside Panicum
Panicum amarum Ell.

[367]
Glaucous glabrous perennial 0.2–3 m tall with extensive rhizomes. Lower stem widths vary from about 2 to 10 mm. Stems solitary to more frequently in clumps. Spikelets 4–7.7 mm long in panicles 10–40 cm long and 2–10 cm wide, the principal branches ascending. Common. Active dune areas, beaches, swales behind foredunes; TX–CT. July–Oct.

Two varieties have been recognized: var. *amarum* with sparsely flowered panicles, first glume with 7 or 9 veins, stems often scattered. The other, var. *amarulum* (Hitchc. & Chase) P. G. Palmer, has densely flowered panicles, first glume with 3 or 5 veins, stems often bunched.

Flat-stemmed Panic Grass
Panicum rhizomatum Hitchc. & Chase

[368]
Perennial to 75 cm tall. Stems solitary or in small tufts, from prominent creeping rhizomes. Spikelets 2.2–2.8 mm long, on short pedicels, mostly situated on one side of the panicle branches. Common. Usually moist or poorly drained soils—thin woods, ditches, thin pinelands; TX–MD. Aug.–Oct.

P. anceps Michx. is similar but larger, to 1 m tall, stems usually in loose clumps, and spikelets 2.8–3.8 mm long. Occasional. Similar places; also in old fields, waste places; TX–NJ. Aug.–Oct.

Panicum rhizomatum

Fall Panicum
Panicum dichotomiflorum Michx.

[369]
Glabrous somewhat succulent annual with geniculate stems, especially at base. Highly variable in size, from about 15 to 100 cm tall; usually much-branched, the larger plants often becoming top-heavy and falling over, forming dense masses. Spikelets about 2.5 mm long in thin panicles. First glume of spikelet about ¼ the length of the spikelet. Common. Dry to more often wet places in the open—depressions, ditches, fields, shrub and flower borders, marshes, swales, sloughs; TX–sNS. Aug.–frost.

Panicum dichotomiflorum

Cupscale
Sacciolepis striata (L.) Nash

[370]
Glabrous perennial with much-branched stems that are often decumbent and rooting at the nodes or even falling over and growing over 2 m long, and often forming dense masses. Spikelets 3.5–5 mm long, much like those of *Paspalum, Panicum,* and *Echinochloa* species, the inner unit being hard, shiny, and smooth. Easily recognized by its inflorescence, which is a slender panicle, the branches so short that it usually is spikelike, and by the unique second glume of the spikelet. This glume is about as long as the spikelet and has a cupped or bulging base appearing much like baggy knees. The first glume is short and acute at apex. Important food source for wildlife. Common. Marshes, swales, sloughs, ditches, pond margins, depressions; eTX–NJ. July–frost.

S. indica (L.) Chase, an annual, has similar but smaller spikelets, 2.5–3 mm long, and the plants are smaller. Rare. Similar places; FL–NC. July–frost.

Sacciolepis striata

Water Grass
Echinochloa walteri (Pursh) Heller

[371]
Annual to 2 m tall with 1 to several usually erect stems from the base, each bearing a terminal inflorescence. Other inflorescences, usually smaller, often occur on erect lateral branches. Inflorescences are panicles that are usually nodding; the branches spikelike with the spikelets congested in short secondary branches. Spikelets lanceolate; about 3 times as long as

Echinochloa walteri

broad; with the inner unit (the fertile lemma) hard, shiny, and smooth, as in *Paspalum, Panicum,* and *Sacciolepis* species. The tip of the fertile lemma is abruptly pointed and does not enclose the tip of the palea. The second glume and sterile lemma are long-awned. Common. Wet places or in shallow water in thin woods or in the open—marshes, ponds, sloughs, depressions, ditches; TX–sMA. July–frost.

E. crusgalli (L.) Beauv. (Barnyard Grass) is similar but spikelets ovate, about twice as long as broad, and awnless or awns under 10 mm long. Common. Dry to wet places in the open—waste places, fields, roadsides, marshes, depressions; TX–NS. July–frost.

Yellow Bristlegrass
Setaria glauca (L.) Beauv.

[372]
Setaria species are conspicuous because of their bristly spikelike panicles. These are distinctive in that 1 to several bristles are attached just under the base of each spikelet, and the spikelet or most of it falls free, leaving the bristles. Spikelets are much like those of *Paspalum* and *Panicum* species, having a hard shiny inner unit, the fertile lemma, which is

tightly clasped by the membranous infertile lemma and 2 glumes. Fruits are an important source of food for wildlife.

S. glauca is an annual to 1 m tall, branching and often forming dense colonies. Leaves mostly spreading, the sheath margins lacking cilia. Panicles 4–12 cm long. Spikelets 2.8–3 mm long, the fertile lemma horizontally finely ridged. Bristles 4–8 below each spikelet. Common. Dry to wet places in the open—roadsides, fields, stable dune areas, around buildings, shrub borders; TX–NS. July–frost. Syn.: *S. lutescens* (Weigel) F. T. Hubb.

Setaria glauca

Knotroot Bristlegrass
Setaria geniculata (Lam.) Beauv.

[373]
Perennial to 80 cm tall with knotty, usually branching rhizomes. Stems generally several from the base and often with branches above. Leaves mostly erect, the sheath margins lacking cilia. Panicles 2–8 cm long. Spikelets 2–2.5 mm long, the fertile lemma finely glanular to very finely roughened. Bristles 4–8 below each spikelet. Common. Thin woods, marsh margins, swales, wet pinelands, pond shores, roadsides, fields, yard borders, fencerows; TX–sMA. May–frost.

Green Bristlegrass
Setaria viridis (L.) Beauv.

[374]
Annual to 80 cm tall, stems branching from or near the base. Leaves mostly erect, the sheath margins ciliate. Panicles 2–10 cm long. Spikelets 1.8–2.5 mm long, the fertile lemma minutely roughened. Bristles 1–3 below each spikelet, green until spikelets are falling. Occasional. Roadsides, disturbed soils, yard margins, gardens, parking

areas; nNC–Nfld. June–frost.

S. corrugata (Ell.) Schult. has a similar aspect and also is an annual but the fertile lemmas are clearly horizontally ridged and the bristles purplish. Occasional. Roadsides, fields, waste places, parking areas; GA–sNC. July–frost.

Giant-millet; Giant Foxtail
Setaria magna Griseb.

[375]
Coarse annual to 4 m tall, stems to 2.5 cm thick at base. Inflorescence to 50 cm long and 6 cm broad, arching to nodding. Spikelets 2–2.2 mm long, the fertile lemma smooth and falling free when mature, leaving the remainder of the spikelet with the bristles. Common. Freshwater and brackish marshes, pond margins, swales, sloughs, ditches; LA–sNJ. July–frost.

S. italica (L.) Beauv. (Italian Millet) has a similar aspect but smaller, to 1 m tall, inflorescence to 30 cm long, and spikelets 2.5–3 mm long. Rare. Roadsides, waste places, fencerows; VA–NS. July–Oct.

Dune Sandbur
Cenchrus tribuloides L.

[376]
Seaside sandburs are easy to recognize; 1–8 spikelets are enclosed in a spiny bur, the spines minutely retrorsely barbed near the tip. Plants often not noticed until burs are clinging to clothes or skin. Parts of spines can break off under the skin and become quite painful.

Dune Sandbur, an annual, is easily recognized by the densely hairy cuplike burs that, including spines, are 10–15 mm wide. Burs with 15–43 spines, in spikelike racemes, the pedicels swollen. Leaf sheaths inflated, blades usually folded. Common. Dunes, loose sands in thin woods or in the open; MS–NY. Aug.–Sept.

Southern Sandbur
Cenchrus echinatus L.

[377]
Tufted annual with ascending to sprawling branched stems. Burs with a ring of bristles at base, these terete and nearly as long as but more slender than the larger-based flattened spines above them. Common. Roadsides, lawns, fields, stable dune areas; LA–SC. June–frost.

Coastal Sandbur
Cenchrus incertus M. A. Curtis

[378]
Tufted annual with short rhizomes, sometimes overwintering. Stems branching and often sprawling. Leaf blades 2–6 mm wide. Burs, including spines, to 7 mm wide, glabrous or with many very short hairs; spines 8–40, broad at base. Ring of basal bristles lacking. Common. Stable dune areas, minidunes, roadsides, waste places, thin woods; TX–sNC. July–frost. Syn.: *C. pauciflorus* Benth.

C. longispinus (Hack.) Fern. may be separated from other seaside sandburs by having 50 or more slender spines, those at the base of the bur numerous, pointing downward, and shorter than those on the body of the bur. Common. Dunes, sandy meadows, roadsides, fields; nNC–MA. July–Oct.

St. Augustine Grass
Stenotaphrum secundatum (Walt.) Kuntze

[379]
Low mat-forming stoloniferous perennial having thick glabrous succulent flat leaf blades with an obtuse apex. Leaf sheaths keeled, loosely clasping stem. Inflorescence a spike, the spikelets partially embedded in the thick flattened axis. Spikes 6–9 mm long, 5–8 mm broad. Common. Commonly cultivated in yards, roadsides, pastures, swales, meadows, brackish marshes, above drift zone; TX–sNC. July–Oct.

Annual Wildrice
Zizania aquatica L.

[380]
Coarse plants to 3 m tall. Mostly perennial southward and annual northward. Principal leaves 4–5 cm wide, to 1.2 m long, margins finely toothed. Inflorescence a large terminal panicle. Spikelets unisexual, the sexes in the same panicle, the male ones hanging from the spreading lower branches of the panicle, the female ones erect on the stiffly erect upper branches. One of the few grasses with 6 stamens per flower. Common. Freshwater and brackish marshes; LA–NBr. May–Oct.

Zizaniopsis miliacea (Michx.) Doell & Asch. (Southern Wildrice) is much the same vegetatively except it is always a perennial. The spikelets are also unisexual and in large panicles but the sexes are intermixed throughout the panicle and all branches of the panicle are spreading. Common. Similar places; TX–MD. Apr.–Aug.

Pink Muhlenbergia; Pink Muhly
Muhlenbergia filipes M. A. Curtis

[381]
Muhlenbergia is one of the many genera with a single floret per spikelet and spikelets in panicles. Identification to genus and species is often difficult. In a few species there are unusual or even unique characters that set them aside.

Muhlenbergia filipes

M. filipes is easily recognized. It is a tufted perennial to 80 cm tall, the inflorescence a loose limber panicle ⅓ to ½ as long as the entire plant, pinkish when mature, or rarely white. When these panicles are in full color and bending with the wind they are outstandingly attractive, especially when there are many clumps over a large area. The floret has a delicate awn 5–15 mm long. The stems are collected along the GA and SC coasts and used as one of the components in the weaving of attractive and serviceable baskets, table mats, and similar items. Common. Moist places in thin woods or in the open—flats between dunes, pinelands; TX–NS. Sept.–Oct.

Coastal Dropseed
Sporobolus virginicus (L.) Kunth.

[382]
Sporobolus is one of many genera with 1 floret per spikelet and spikelets in panicles, but is fairly easy to recognize when fruits are mature, as it is one of 2 seaside genera having seeds free from the ovary wall. This wall is thin and closely covers the seed. Rolling mature fruits between thumb and finger will remove some of the ovary wall, revealing the shiny and reddish seed.

Coastal Dropseed is a perennial to 40 cm tall from extensive rhizomes forming dense colonies often covering large areas. Leaf sheaths overlap. There are 2 forms: a small one in salt or brackish marshes, swales, or overwash areas with leaves about 5 cm long or less. The other, a robust form, occurs mostly in accretion areas along beaches and on dunes, especially dunelets. Leaves of these plants grow to 17 cm long and 7 mm wide. Stems, rhizomes, and panicles are also larger. As may be seen in the photograph, the panicle is spikelike. It is easily confused with that of the common seashore salt grass. Careful examination is usually needed to find the ultimate branches of the inflorescence and determine that the spikelet on the end has only 1 floret. The floret is clasped by 2 glumes about as long as the floret. Spikelets 2.3–3.4 mm long. Common. Salt or brackish marshes, overwash areas, swales, dunes, minidunes, salt flats; TX–sNC. June–Nov.

Sporobolus virginicus

Smut Grass
Sporobolus indicus (L.) R. Br.

[383]
Tufted perennial to 80 cm tall with mostly basal leaves and a slender panicle to 40 cm long. Branches of the panicle are often conspicuous though closely ascending and the lower 1 to few occasionally not overlapping those immediately above. Leaves are 2–5 mm wide and quite tough, difficult to cut with mowers. Healthy panicles are greenish when mature, turning light tan upon drying. Seeds at maturity are reddish and loose, tending to stick to the mucilaginous ovary wall from which they came. Panicles are often infected with a smut (a fungus), which turns them black. Common. Meadows between dunes, swales, thin maritime or live oak woods, thin pinelands, lawns, pathways, roadsides; eTX–VA. Mar.–Nov. Syn.: *S. poiretii* (R. & S.) Hitchc.

Rabbitfoot Grass
Polypogon monspeliensis (L.) Desf.

[384]
Tufted annual to 70 cm tall, often diminutive in poor habitats. Spikelets in dense panicles, the glumes with awns 8–15 mm long that are a conspicuous part of the panicles. Spikelets fall in their entirety, the single floret (smooth, shiny, and about 1 mm long) hidden by the clasping

glumes except sometimes for the tip of the delicate awn on the floret. The upper margins of the glumes are entire. Especially conspicuous in the morning dew. Occasional. Fresh to brackish places in the open—ditches, swales, pond margins, depressions, marshes; TX–VA. Mar.–July.

P. maritimus Willd. is nearly identical but the upper margins of the glumes are ciliate. Rare. Similar places; GA–NC. Apr.–June.

Polypogon monspeliensis

Redtop
Agrostis stolonifera L.

[385]
Inflorescences of *Agrostis* species are loose panicles. Spikelets have 2 glumes equal or longer than the single floret. The glumes and lemma are thin. Awns are lacking or sometimes a short one present on back of the floret, not at the tip. Florets fall from the panicle at maturity, leaving the glumes.

Redtop is separated from other species of *Agrostis* by having spikelets borne nearly to the base of the panicle branches. Spikelets are 2–3 mm long and individually pedicelled. Common. Open usually moist to wet places—freshwater to brackish marshes, meadows, pond edges, fields, roadsides, swales; NC–Greenl. June–Oct. Syn.: *A. alba* L. as used in some manuals.

In three similar seaside species spikelets are not borne near the base of the panicle branches. In *A. perennans* (Walt.) Tuckerm. (Autumn Bent-

Agrostis stolonifera

grass) the innermost panicle branches are at or just below the middle. Spikelets are 2–2.5 mm long and scattered along most of the secondary branches. Common. Similar places; NC–NS. July–Oct. In the other 2 species the branches fork well beyond the middle and the spikelets are distributed near the ends of the ultimate branches. Spikelets are 1.5–1.8 mm long in *A. hiemalis* (Walt.) B.S.P. (Ticklegrass). Common. Roadsides, old fields, meadows, pond shores; MS–MA. Mar.–June. In *A. scabra* Willd. spikelets are 2.0–2.7 mm long. Occasional. Places similar to Ticklegrass; GA–Greenl. June–Nov.

American Beachgrass
Ammophila breviligulata Fern.

[386]
Rhizomatous perennial to 1.5 m tall. Most leaves are basal. Inflorescence a dense spikelike panicle 15–40 cm long and 10–25 mm broad. Spikelets 10–15 mm long, with 1 floret, glumes and lemmas awnless. Lemma with a tuft of short hairs at base and falling from spikelet at maturity, leaving glumes. Important as a sand-binder. Common. Dunes, loose sands, sandy beaches; NC–Nfld. July–Sept.

A. arenaria (L.) Link (European Beachgrass) has been introduced at Cape Cod, MA, and is spreading. Its panicle is lanceolate and 10–20 cm

Ammophila breviligulata

long. The ligule is thin and 10–30 mm long whereas that of American Beachgrass is firm and 1–3 mm long. Rare. Similar places; MA. July–Aug.

Velvet Grass
Holcus lanatus L.

[387]
Tufted annual to 1 m tall, the entire plant grayish velvety-hairy and soft to touch. Glumes 3.2–4.5 mm long, ciliate on the keeled back. Florets 2; the lemmas 1.8–2.0 mm long, the upper one with a hooked awn 1–2 mm long. Common. Fields, roadsides, waste places, meadows; thin woods; NC–NS. May–Sept.

Big Cordgrass
Spartina cynosuroides (L.) Roth

[388]
Spartina is the most important genus in seaside habitats. Only 6 species, all perennials, are involved but the number of plants is astronomical. The genus is fairly easily recognized. Spikelets are arranged in 2 rows on 1-sided spikes and these distributed along a central axis in numbers ranging from 3 to 75. Spikelets consist of 2 glumes and 1 floret, its palea usually being evident as well as the lemma. The lemma is keeled and of the same general texture as the glumes and not at all similar to the hard, shiny, and round-backed floret of the spikelets in the 1-sided spikes of *Paspalum* species.

Big Cordgrass grows to 3.5 m tall. Stems hard. Fresh leaf blades flat, the largest 10–25 mm wide, the margins scabrous. Panicle with 5–67 spikes, these spreading. Longer of the 2 glumes ⅓ longer than the floret. Common. Brackish or freshwater tidal marshes, brackish sloughs; eTX–sMA. June–Oct.

S. pectinata Link (Prairie Cordgrass), which grows to 2.5 m tall, also has hard stems, fresh leaves flat, these 5–15 mm wide and with scabrous margins. Spikes vary from 5 to 50 and are loosely appressed to spreading. The longer of the 2 glumes

Spartina cynosuroides

is about twice as long as the floret, longer than in any other species. Common. Freshwater and saltwater habitats—marshes, shores, ditches, pond shores; NJ–NS. June–Sept.

Smooth Cordgrass
Spartina alterniflora Loisel.

[389]
This is the most important species in seaside habitats, covering vast areas to the exclusion or nearly so of other species. Plants grow to 2.5 m tall and have soft stems. Leaf blades are flat, not scabrous on the margins, and 3–25 mm wide. Panicles erect to arching. Spikes are 3–25, loosely appressed, separated or more frequently moderately overlapping. Plants in the high salt marshes, especially at edges of salt pans, may be only 40 cm tall including the inflorescence. Mid to high tide levels in salt and brackish marshes; TX–Nfld. June–Oct.

S. spartinae (Trin.) Merr. (Gulf Cordgrass) is easily recognized. It grows to 2 m tall in clumps to 70 cm in diameter. Leaf blades are involute when fresh and only to 5 mm wide. Spikes are 6–75, tightly appressed, and closely overlapping in panicles 6–70 cm long. Common. Dunes, sandy beaches, roadsides, ditches, meadows, salt flats, marshes; TX–wFL. Mar.–Oct.

Marshhay; Saltmeadow Cordgrass; Salthay
Spartina patens (Ait.) Muhl.

[390]
Plants to 1.5 m tall, solitary or in small clumps from widely spreading slender wiry rhizomes. Leaf blades

involute or rarely flat, 0.5–5 mm wide. Spikes 2–15, alternate, often well separated, spreading to less commonly loosely appressed. Longer glume blunt-pointed, about ⅓ longer than the lemma. Once abundantly cut for hay and still harvested in some northern areas. Common. High salt marsh, salt meadows, brackish flats, beaches, overwash areas, low dunes, swales; TX–NS. June–Sept.

S. *bakeri* Merr. (Bunch Cordgrass) is similar in many respects but rhizomes are lacking and plants are in circular colonies to 1 m across. The longer glume is sharply tapered to a fine point. Common. Swales, depressions, pond margins, at edge of brackish marshes, wet pinelands, meadows; FL–sSC. Dec.–Mar., occasionally in summer.

Fingergrass
Eustachys petraea (Sw.) Desv.

[391]
Tufted perennial to 60 cm tall, with short rhizomes, stems often decumbent. Leaves mostly basal, glabrous, apex rounded to broadly acute. Spikelets many, 1.8–2.2 mm long, closely arranged on 2 sides of a 3-sided spike. Spikes 3–6, to 6 cm long, arising about the same point at end of stem. Axis of spikes extending slightly beyond the spikelets. Fertile lemma shiny brown, at maturity falling and leaving the glumes, the spikes then appearing feathery. Common. Swales, sandy flats, roadsides, grasslands, margins of freshwater and brackish marshes, thin live oak woods; eTX–NY. June–Oct. Syn.: *Chloris petraea* Sw.

E. *glauca* Chapm. is larger, to 1.2 m tall, and has 8–16 ascending to erect spikes 8–20 cm long. Rare. Margins of freshwater and brackish marshes, depressions, ditches; AL–sNC. June–Oct. Syn.: *Chloris glauca* (Chapm.) Wood.

Goosegrass
Eleusine indica (L.) Gaertn.

[392]
Coarse tufted annual with mostly decumbent stems and overlapping leaf sheaths. Leaves and stems tough. Spikelets many, arranged in 2 rows on one side of spikes, 3- to 6-flowered, lemmas and glumes awnless and essentially the same texture. Spikes clustered at tip of stem or occasionally 1 below, the axis not extending beyond the spikelets. So tough that goosegrass is the last to disappear in well-used footpaths or vehicle ways. Common. Footpaths, vehicle trails, lawns, parking lots, around buildings, waste places; eTX–NY. June–Oct.

Eleusine indica

Purple Sandgrass
Triplasis purpurea (Walt.) Chapm.

[393]
Tufted annual with wiry stems to 1 m long. Stems erect or more often decumbent or nearly prostrate. Inflorescence only 3–5 cm long with few spikelets. Spikelets with the 2 glumes smaller than the 2–4 lemmas, which are densely short-hairy on the midrib and the 2 lateral veins and have an awn under 3 mm long on the tip. Common. Dunes, especially in places devoid of other plants or nearly so; dry sandy flats; thin maritime, live

Triplasis purpurea

oak, and pine woods; TX–sME. Aug.–Oct.

T. americana Beauv. (Perennial Sandgrass) is quite similar but stems are thinner and not as long, and the awn on the lemmas is 5–8 mm long. Common. Similar places but infrequent on active dunes; MS–sNC. Sept.–Oct.

Common Reed
Phragmites australis (Cav.) Trin. ex Steud.

[394]
Coarse rhizomatous perennial to 4 m tall, forming extensive dense colonies. Leaf blades to 50 cm long and 5 cm broad, the sheaths overlapping. Inflorescence a dense tawny to purplish panicle to 50 cm long and 20 cm wide. Spikelets with thin glumes and lemmas. Glumes 2, shorter than the 3–8 loosely arranged glabrous lemmas. The conspicuous hairs in the spikelets arise from the axis. Common. Marshes, especially tidal; pond margins; ditches; disturbed habitats; TX–NS. July–Oct. Syn.: *P. communis* (L.) Trin.

Arundo donax L. (Giant Reed) is nearly identical except to 6 m tall, the glumes longer than the lemmas, and hairs in the spikelets arising from the base of the lemmas. Planted and persisting or spreading vegetatively in dense colonies by means of short thick rhizomes. Rare. MS–Va. Sept.–Oct.

Purple Lovegrass
Eragrostis spectabilis (Pursh) Steud.

[395]
A large genus with inflorescences of panicles varying from large, open, and thin to small and dense. Spikelets with glumes and lemmas of similar texture, glumes shorter than the largest lemma, florets 2 to many. Most plants are difficult to name. We have recorded 10 species in seaside habitats; only 3 are common. They are similar and are included.

E. spectabilis is a tufted perennial to 80 cm tall with stiffly erect to spreading stems. Each panicle is usually about ⅔ of the height of the plant, and has branches spreading to ascending near the top. The whole panicle eventually breaks away and tumbles before the wind. Spikelets 4–8 mm long, usually purplish, on slim but rigid pedicels longer than the spikelets. Florets 3–10. Common. Stable dune areas, loose sands, fencerows, fields, dry pinelands, thin live oak woods; eTX–ME. July–Oct.

Two similar species have flexible capillary pedicels. In *E. refracta* (Muhl.) Scribn. the pedicels are longer than the spikelets, which, except for terminal ones, are pressed against the inflorescence branches. Common. Marshes, pond margins, ditches, depressions; eTX–DE. July–Oct. In *E. elliottii* S. Wats. the spikelets are longer than their pedicels and are divergent. Common. Similar places; eTX–NC, but rare along the Atlantic Ocean. Aug.–Nov.

Eragrostis spectabilis

Sea-oats
Uniola paniculata L.

[396]
Coarse perennial to 2 m tall with extensive creeping rhizomes, readily rooting at nodes when covered with sand; the rhizomes, roots, leaves, and stems are highly important in stabiliz-

Uniola paniculata

ing dunes. Inflorescence a panicle. Spikelets conspicuously flattened, with 8–20 florets. In most states, as a preservation measure, it is unlawful to gather any part of the plant. Common. Dunes, beaches, loose sands near seashores; TX–VA. June–Sept.

Spangle Grass
Chasmanthium sessiliflorum (Poir.) Yates

[397]
Perennial to 1.5 m tall with short rhizomes. Often in thin extensive colonies. Stems tufted, suberect to arching. Leaves hairy at junction of sheath and blade, the largest leaves occurring a little above the base. Inflorescence a narrow panicle, the branches stiffly ascending and appressed or less commonly spreading as in the photograph. Spikelets with glumes 1–2.5 mm long. Larger lemmas acuminate and somewhat divergent, the shorter palea usually evident between the lemma and the axis of the spikelet. Common. Usually in drier soils—live oak, mixed deciduous, pine-broadleaf woods; eTX–sVA. July–Sept. Syn.: *Uniola sessiliflora* Poir.

C. *laxum* (L.) Yates is very similar but may be recognized by absence of hairs at junction of sheath and blade. Common. Usually in moist soils of similar habitats and in meadows and swamps; eTX–VA. June–Oct. Syn.: *Uniola laxa* (L.) B.S.P.

Saltgrass
Distichlis spicata (L.) Greene

[398]
Perennial to 75 cm tall, usually under 40 cm, with extensive rhizomes. Plants in dense colonies or in lines as stems arise from the rhizomes. Leaves stiff, diverging about 45° from the stem, obviously 2-ranked, sheaths overlapping. Inflorescence a dense panicle 2–7 cm long, sometimes only a few spikelets with pedicels. Spikelets mostly with 5–9 florets, these unisexual, the sexes on different plants. Saltgrass and Coastal Dropseed, which often grow together or close

by, are not distinguishable vegetatively. The spikelets, however, are radically different; the latter species has 2 glumes and 1 floret, whereas in Saltgrass there are 2 glumes and 5–9 florets. Careful examination is often needed to find the ultimate branches of the inflorescence to recognize the flattened spikelet with its 5–9 2-ranked florets. One of the most important plants in salt marshes. Common. Higher portions of salt and brackish marshes, active overwash areas, accretion areas, brackish or salt pools and ponds, salt flats; eTX–NS. June–Oct.

Distichlis spicata

Quaking Grass
Briza minor L.

[399]
The spikelets of *Briza* species are so different from those of other grasses that they readily attract attention. They are deltoid, hang down on thin curved pedicels, and are easily moved by wind. Some details of the spikelet are also peculiar; the lemmas are almost horizontal, have thin broad margins and a rounded apex.

B. *minor* is an erect glabrous annual to 70 cm tall. Inflorescence an oblong to pyramidal panicle 2–15 cm long and nearly as wide. Spikelets are about 3 mm long and a little wider. Occasional. Swales, fields, waste places, thin woods; eTX–sVA. Mar.–May.

Annual Bluegrass
Poa annua L.

[400]
Plants to 40 cm long, but usually under 15 cm; bright green; tufted; often forming dense mats, excluding

other plants; dying in late spring and leaving light-colored mats, these unsightly and therefore troublesome in lawns. Inflorescence a pyramidal panicle 1–12 cm long. Spikelets with glumes and lemmas alike, lemmas 3–6. Common. Lawns, swales, paths, roadsides, waste places, parking lots, crevices of sidewalks, shrub and flower plantings; eTX–Greenl. Mar.–May.

Annual Fescue
Vulpia octoflora (Walt.) Rydb.

[401]
Slender erect tufted annual 3–60 cm tall. Leaves basal and on lower part of stem; blades to 8 cm long, very narrow, under 1 mm wide. Inflorescence a narrow panicle, the larger ones with divergent branches but with appressed spikelets. Spikelets with nearly equal glumes, these shorter than the largest lemma. Florets 5–13, the lemmas glabrous or scabrous, usually awned, the awns to about 7 mm long. Often in loose colonies, these especially conspicuous. Common. Swales, yard margins, roadsides, fields, shrub and flower borders, various poor and loose soils in open; TX–sME. Mar.–June. Syn.: *Festuca octoflora* Walt.

V. sciurea (Nutt.) Henr. is much similar but awn on the lemmas is 2–4 times as long as the body, which has small scattered hairs. One glume about ½ as long as the other one. Occasional. Similar places; TX–sNJ. Mar.–June. Syn.: *Festuca sciurea* Nutt. In the similar *V. myuros* (L.) K. C. Gmel. the shorter glume is ¼ to ⅓

the length of the other one and the awns on the lemmas are 2–5 times as long as the body. Occasional. Similar places; TX–sME. Apr.–June. Syn.: *Festuca myuros* L.

Wild Ryegrass
Elymus virginicus L.

[402]
Elymus species are erect tufted perennials with short rhizomes. Spikelets are sessile, arranged in groups (usually 2 each) alternately on 2 sides of a spike. This arrangement is most apparent at the base of the spike or if the spike is bent. Glumes and lemmas are awned or rarely blunt or merely acute. The glumes are not on opposite sides of the spikelet but are side by side so that if there are 2 spikelets there is an arc of 4 glumes; if 3 spikelets, an arc of 6 glumes.

In *E. virginicus* the largest glume is 1.5–2 mm broad at widest part and the awns are nearly straight. Common. In thin woods or in the open—depressions, meadows, shores of freshwater marshes, swales, shell mounds, fields; eTX–Nfld. June–Oct.

E. villosus Muhl. is similar but glumes are under 1 mm wide. Rare. Similar places; nNJ–NH. June–Aug.

Elymus virginicus

Vulpia octoflora

CYPERACEAE
Sedge Family

Sedges are common constituents of most seaside habitats, sometimes abundantly so. Some are easy to identify, many require a specialist. We are including one to several common representatives of nearly all genera, believing that they can be identified at least to genus and most to species with reasonable confidence.

Most sedges resemble grasses vegetatively and are often confused with them. They are easily distinguished, however, even when not in flower or fruit. The characters used for this purpose are included among those given in the introductory section to grasses (p. 323). The most important characteristics of sedges can be summarized as follows: leaves 3-ranked, internodes usually solid and usually triangular in cross section, margins of leaf sheaths fused, flowers (florets) single in axil of each bract (also called scale) or the lower 1 or 2 scales sometimes empty. These scales are in either 2-ranked or tightly spiralled groups called spikelets. Fruit is an achene or nutlet. Sedges are frequent components of dried flower arrangements.

Yellow Nutgrass
Cyperus esculentus L.

[403]
Cyperus is one of the easiest genera to recognize. The inflorescences are terminal and the florets are 2-ranked. The spikelets are therefore very much like those of grasses. *Cyperus* species are easily distinguished from grasses, however, by their 3-ranked leaves. Also, all scales bear flowers; there are no empty glumes at the base as in almost all grasses. We have recorded 33 species in seaside habitats. Their identification is usually difficult so we include only a few of the most common ones, our choices designed in part to provide some idea of variation within the genus.

Yellow Nutgrass is a perennial to 70 cm tall with slender rhizomes that bear hard tubers. Leaves mostly basal and well developed. Spikelets 8- to 30-flowered, straw-colored to golden-brown, individually quite evident, numerous, and mostly horizontally attached in spikes. Spikes a few to 30 or more, the spike axis visible. Scales and achenes shedding at maturity leaving the spikelet axis, which falls later. Achenes 3-angled. A troublesome weed in many places. Common. Beach sands, dunes, swales, meadows, roadsides, parking areas, paths, yards, gardens, fields; TX–NS. June–Nov.

Flatsedge
Cyperus strigosus L.

[404]
Perennial 10–100 cm tall, usually tufted, with a hard swollen base. Leaves largely basal, conspicuous. Inflorescence with a few conspicuous leaflike bracts at base. Spikelets similar to and arranged much like those of yellow nutgrass but with a few to many spikelets that are not horizontally arranged, the spike axis visible. The entire mature spikelet falls from axis at maturity. Achenes 3-angled. Common. Swales, pond margins, ditches, depressions, roadsides, marsh margins; eTX–sMA. June–Oct.

C. odoratus L. has a similar aspect and is easily recognized because at maturity the axis of the spikelets breaks into joints that fall with scales and achenes attached. Achenes are 1–2 mm long and 3-angled. Common. Similar places; TX–MA. July–Oct.

Cyperus strigosus

Leafless Sedge
Cyperus haspan L.

[405]
Short-lived tufted perennial 20–70 cm tall, flowering first year. Stem sharply angled, soft, weak. Leaf blades essentially missing, the sheaths enlarged and conspicuous. Leaflike bracts at base of inflorescence 1 to several, not conspicuous, one equaling or a little longer than the inflorescence, any others much smaller. Spikelets in umbel-like groups, shedding scales and achenes as they mature, leaving the axis. Achenes 3-angled. Common. Often in shallow water, swales, depressions, sloughs, pond margins, ditches, freshwater and brackish marshes; TX–sVA. June–Oct.

Flatsedge
Cyperus retrorsus Chapm.

[406]
Perennial 20–100 cm tall, the base thickened and usually hard. Leaves basal and on lower part of stem. Bracts at base of inflorescence 3–8, variable in size, one much exceeding the inflorescence. Spikes are few to numerous, dense, cylindrical; they have many spirally arranged spikelets—the uppermost erect or nearly so, the middle ones about at right angles, and the lower ones reflexed. Spikelets 2–6 mm long with 1–4 florets. Achenes 3-angled. Common. Dry to wet places—swales, margins of fresh and brackish marshes, meadows, pond margins, ditches, fields, pinelands; eTX–NY. July–Oct.

In *C. globulosus* Aubl. the spikes are globose, the spikelets with 3–6 florets and radiating outward in all directions. Perhaps not distinct from *C. retrorsus*. Common. Similar places; eTX–VA. July–Oct.

Cyperus retrorsus

Umbrella Sedge
Cyperus filicinus Vahl

[407]
Slim tufted perennial to 40 cm tall but sometimes flowering when only 3 cm tall, often the first year. Leaves mostly basal, to 2 mm wide. Bracts under the inflorescence 2–3, to 2.5 mm wide. Spikelets many-flowered, the scales 2.5–3 mm long. Spikes 1–6, roughly globose. Achenes 1–1.5 mm long, lens-shaped, linear-oblong to narrowly oblong-ovate; as they near maturity they become brown, at maturity grayish-iridescent. Common. Swales, ditches, wet pinelands, pond margins, upper parts of brackish marshes, freshwater marshes, depressions; eTX–sME. July–Oct.

C. polystachyos Rottb. var. *texensis* (Torr.) Fern. is closely similar but the scales are only 1.5–2 mm long and the achenes 0.8–1 mm long. Considered by some as not being separable from *C. filicinus*. Common. Similar places; eTX–MA. July–Oct.

One-headed Flatsedge
Cyperus brevifolius (Rottb) Endl. ex Hassk.

[408]
Rhizomatous perennial to 40 cm tall, often forming mats. Lower leaves with sheaths only; the others mostly near the base, 1–3 mm wide, 1–10 cm long. Bracts below the inflorescence 3–4, unequal, to 25 cm long. Inflorescence a single head, a spike, 3.5–12 mm long, containing many 1-flowered spikelets 2–3 mm long. Each spikelet produces only 1 achene. These are lens-shaped and oblong to obovate and 1–1.3 mm long. Occasional. Moist to wet places—swales, thin pinelands, ditches; eTX–GA. Apr.–Nov. Syn.: *Kyllinga brevifolia* Rottb.

Umbrella-grass
Fuirena pumila (Torr.) Spreng.

[409]
Fuirena species have obtusely triangular stems and the lowermost leaves

consist entirely or essentially of sheath only. Spikelets are compact, many-flowered, mostly in tight terminal clusters, sometimes with additional peduncled clusters in 1–3 of the next lowest leaf axils. The flowers are in axils of scales that are densely and spirally arranged. Flowers are bisexual. Perianth small but distinctive, consisting of 3 broad-pedicelled scales alternating with 3 bristles.

F. pumila is a tufted annual to 60 cm tall, mostly 30 cm or less, with soft stems. Leaf blades spreading to ascending to 12 cm long and 5 mm broad, basal margins with spreading hairs. The spikelet scales are about 3 mm long and with a conspicuous awn; the 3 perianth scales behind each of them about as long and with a mostly ovate blade and bristlelike tip. Common. Moist to wet places—swales, pond margins, depressions, swamps; FL–MA. July–Oct.

F. squarrosa Michx. is similar but is a perennial from short scaly rhizomes, sometimes taller, to 1 m, and leaf blades to 20 cm long and 10 mm broad. Common. Similar places; MS–wFL; SC–NY. July–Oct. Syn.: *F. hispida* Ell.

Leafless Fuirena
Fuirena scirpoidea Michx.

[410]
Erect glabrous perennial to 60 cm tall from well-developed rhizomes just beneath the soil surface, the stems often lined up like parts of a picket fence. Leaves mostly of sheaths; the blade, if any, cuplike, erect, to 4 mm long. Spikelets ovoid to lanceolate-ovoid, 7–12 mm long, 1–5 of them sessile in a terminal cluster; the bract at their base shorter than the spikelets. Common. Swales, depressions, edge of brackish marshes, pond margins, ditches, wet pinelands; MS–FL. Mar.–Aug.

F. longa Chapm. is similar but upper stem leaves with blades to 5 cm long and spikelets shorter than the bract below them. Rare. Similar places; MS–FL. May–July.

Swordgrass
Scirpus americanus Pers.

[411]
Seaside *Scirpus* species are erect perennials with unbranched stems. Spikelets are 1 to many in terminal inflorescences, sometimes appearing lateral when there is only 1 leaflike bract, which appears as an extension of the stem (as in *S. americanus*). Flowers hidden by scalelike bracts that are compactly and spirally arranged. Perianth represented by bristles that may be inconspicuous, even if observed with a hand lens, or relatively evident when the bristles are long and project beyond the floral scales. Fruit an achene. We have recorded 19 species in seaside habitats but since identification of most is difficult we include only 4, showing some of the variations in the genus.

Swordgrass has sharply triangular stems growing to 1.5 m tall and arising from coarse reddish branching rhizomes. Lower leaves with sheaths only. Above these are a few leaves with flat blades to 20 cm long. Spikelets 1–8, sessile, 5–20 mm long, the scales with a conspicuous awn. Involucral bract 1, triangular, 3–12 cm long, appearing as a continuation of the stem. Achenes about 3 mm long. Common. Freshwater or brackish marshes, swales, shallow brackish water; TX–Nfld. June–Sept.

S. olneyi Gray has very much the same appearance but spikelets 4–15 and scales lacking an awn, having only a sharp abrupt point. Occasional. Similar places, eTX–NS. June–Sept.

Saltmarsh Bulrush
Scirpus robustus Pursh

[412]
Rhizomatous perennial to 1.5 m tall, basal leaves of acuminate sheaths only, those along stem to near inflorescence with blades. Bracts at base of inflorescence 2–4, the largest much like blades of the leaves just below. Spikelets 3 to many, 1–3 cm long, 8–12 mm wide, at least some of them

pedicelled. Common. Higher parts of salt or brackish marshes, swales, depressions, ditches; eTX–NS. July–Oct.

Marsh Bulrush
Scirpus cyperinus (L.) Kunth

[413]
Perennial to 2 m tall, usually in large clumps, often in extensive colonies. Leaves long and conspicuous, especially the basal ones. Inflorescence large, the branches arching and drooping. Bracts at base of inflorescence 2–3, at least 1 quite long. Spikelets about 200–500, ovoid, 3–5 mm long, reddish brown. Perianth bristles 6, curly, much longer than the scales, causing the spikelets to appear wooly. Achene yellowish to whitish, 0.7–1 mm long. Common. Freshwater marshes, pond margins, swamps, wet meadows, sloughs; eTX–Nfld. July–Oct.

Spike-rush
Eleocharis tuberculosa (Michx.) R. & S.

[414]
Spike-rushes are distinctive in having a single terminal spikelet of spirally arranged flowers in axils of scales, no bracts below the spikelet, and leaves consisting of bladeless sheaths only. The perianth consists of 6–9 inconspicuous bristles. Fruits are achenes. Various species range from 1 cm to 1 m tall. Identification of most species is difficult. We have recorded 30 species in seaside habitats but have chosen only 2 fairly easy-to-recognize ones to illustrate and to give an idea of what *Eleocharis* species look like.

E. *tuberculosa* is a tufted perennial to 80 cm tall with fairly stiff erect stems 0.5–1.0 mm thick, somewhat flattened. The achene is triangular,

Eleocharis tuberculosa

obovoid, 1–1.5 mm long, roughly and coarsely honeycombed. On its apex there is a distinctive tubercule as broad and as long as the achene body and capping the achene although almost free from it. Common. Moist to wet places—pond margins, ditches, sloughs, freshwater marshes, depressions in thin woods; eTX–NS. June–Sept.

Spike-rush
Eleocharis flavescens (Poir.) Urban

[415]
This is one of the smaller species, growing from 2–30 cm tall. It is a perennial with very thin rhizomes, forming mats. The upper part of the leaf sheaths is thin, very loose, and transparent. Spikelets 3–6 mm long. The achene is lenticular, smooth, shiny, 0.6–1.0 mm long, purplish-black. The tubercle on the end is minute, conic, about as wide as tall, less than half as wide as the achene. Common. Swales, sloughs, ditches, pond margins, edge of marshes; eTX–sNJ. June–Oct.

E. *olivacea* Torr. is almost identical but may be distinguished by having olive-green to dark-brown achenes. Occasional. Similar places; VA–NS. June–Oct. E. *parvula* (R. & S.) Link ex Buff. & Fingerh. is of similar stature and appearance, and upper part of leaf sheath is thin, loose, and transparent. However, the achene is triangular, obovoid, light-brown, and 0.8–1.5 mm long including the tubercle, which is almost imperceptibly united with the body of the achene. Common. Similar places; TX–Nfld. July–Oct.

Fimbristylis
Fimbristylis castanea (Michx.) Vahl

[416]
This genus cannot be described in a manner useful here but some species are too important to be omitted. Of the 7 species we have recorded for seaside habitats, 4 are rare. Two of the others are especially common, are similar to each other, and are in-

cluded under 1 illustration. These are somewhat similar to some *Scirpus* species but have no perianth bristles, in fact no perianth at all.

F. castanea is a densely tufted perennial to 1.8 m tall and set deeply in the ground. Upper part of stems nearly circular in cross section or at the most elliptical. Leaves usually under 2 mm broad, thick, ascending, mostly basal, rarely any from the upper ⅔ of the stem. Inflorescence with bracts at base, the longest no longer than the inflorescence. Spikelets 5–10 mm long, the scales spiraled. Achenes lenticular, obovoid, 1.5–2.0 mm long. Common. Salt or brackish marshes, swales, sloughs; TX–NY. July–Oct.

F. caroliniana (Lam.) Fern. is similar but rhizomatous, the stems single or in small tufts. Upper part of stems flattened instead of rounded. Common. Beaches, swales, pond shores, ditches; TX–NY. June–Oct.

Whitetop Sedge
Dichromena latifolia Baldw. ex Ell.

[417]
The 7–10 conspicuous flowerlike, white-based leaves (bracts) at the top of these plants positively identify this species. The bracts surround a tight cluster of spikelets with spiral scales. Plants to 1 m tall arising from rhizomes 2–4 mm thick, sometimes forming extensive colonies that can be easily seen from a considerable distance. Occasional. Moist to wet places—roadside flats, depressions, pond margins, wet pinelands; eTX–NC. Apr.–Sept.

D. colorata (L.) Hitchc. has much the same appearance but plants to 60 cm tall and bracts only 4–6. Details of the achenes also separate the species. Common. Swales, ditches, depressions, roadside flats, wet pinelands, pond margins; eTX–sVA. May–Sept.

Saw-grass
Cladium jamaicense Crantz

[418]
Coarse perennial 1–3 m tall from short rhizomes about 1 cm thick,

forming dense impenetrable colonies. Stems obtusely 3-angled. Leaves of lower ⅔ of stem to 1 m long, flat, to 1.5 cm wide, the margins and midrib below with small but dangerous sawteeth. Upper portion of the stem bearing several separated axillary inflorescences arising from axils of relatively small leaves. Spikelets 2–6 on ends of the many branches of the inflorescence, with 3–4 spirally arranged scales, but only 1 achene is formed. Achene 2–2.5 mm long. Perianth absent. Common. Brackish or freshwater marshes; eTX–sVA. July–Oct.

Beaked-rush
Rhynchospora corniculata (Lam.) Gray

[419]
Beaked-rushes are fairly easy to recognize, although magnification higher than provided by a 10 × hand lens may be necessary for species with the smaller flowers. Plants have leafy stems, the leaves with conspicuous blades. Spikelets are numerous in terminal and/or axillary clusters on the upper part of the plant, single spikelets or small clusters of them with evident stalks. Grouped flowers have leaflike bracts at their base. Spikelets have several spiraled scales, but only 1 or 2 bear achenes. Achenes have a tubercule on their tip, which varies from 0.1–20 mm long. Perianth bristles are nearly always present at base of achenes. We have recorded 43 species in seaside habitats. Identification of almost all species is very difficult and cannot be undertaken here. Two species having exceptionally large spikelets can probably be recognized and are included.

R. corniculata is a perennial to 1.5 m tall, usually tufted, with coarse leaves to 2 cm wide. Inflorescence to 55 cm high and 30 cm across. Spikelets lanceolate, about 25 mm long. Achene 3.5–5 mm long; the tubercule long-subulate, 10–20 mm long, protruding prominently beyond scales. Perianth bristles shorter than to as long as achene body. Common. Shallow water and shores of ponds, ditches, depressions, swamps, sloughs; eTX–sVA. June–Oct.

R. macrostachya Torr. has almost identical tubercules but the perianth bristles are longer than the achene body. Also, spikelets are in tight clusters and the inflorescence branches are more erect. Plants are generally smaller, to 1.3 m tall. Occasional. Similar places; eTX–sME. July–Oct.

Nut-rush; Stone-rush
Scleria triglomerata Michx.

[420]
Scleria species are easily recognized by their white or whitish, bony, mature achenes. Stems are sharply 3-angled, the lowest leaves bladeless, the upper with keeled blades. Flowers unisexual, the sexes in the same cluster. Perianth absent. Achene without a tubercle.

S. triglomerata is a perennial with thick knotty rhizomes, often growing in clumps, the stems to 1 m long and spreading or, uncommonly, erect. Blade-bearing leaves mostly in mid ⅓ of stem. Blades to 9 mm broad. Inflorescence terminal accompanied by 2 conspicuous and unequal leaflike bracts, often with 1 to few stalked clusters from axils of leaves immediately below. Achene smooth, shiny, white, 2.5–3.5 mm long, surrounded at base by a thin finely granular disc. Common. Live oak and maritime woods, mixed deciduous woods, pinelands; eTX–NY. May–Oct.

S. reticularis Michx. is similar, a little smaller, stem to 70 cm long; leaf blades only 2–5 mm wide; the achenes dull white or grayish, the surface with a net of small ridges, the base clasped by 3 bractlike structures. Occasional. Moist or wet places—pond shores, depressions, low pinelands, slough margins, low meadows; eFL–sMA. June–Oct.

Sedge; Carex
Carex crinita Lam.

[421]
Carex species have unisexual flowers, the sexes in different spikes or in different parts of the same spike. Both types may be seen in the species illustrated; the terminal spike contains only male flowers, the basal spike only female flowers, and the 3 other spikes with male flowers towards the ends (1 is hidden). The spikes in *Carex* species vary much in size and shape. They are borne singly on stalks as in the photograph, or in dense clusters. If a female flower can be located, mostly by the 2–3 protruding stigmas, or the achenes located, the genus can be positively identified by the presence of the unique saclike structure (the perigynium) surrounding the pistil and later the achene. This is not always easy and identification to species is quite difficult. *Carex* species are important, however, in seaside habitats. From our own field and herbarium work and from published records, we have tabulated over 75 species.

Because it is impractical to include a significant number of the 75 species in a general guide of this type, there is little point to including more than one token species as an example. A common species has been selected. We do not think, however, that any purpose is served by listing the technical features that distinguish it from the other 74 species. Its distribution might be helpful. Common. Depressions, sloughs, swales, pond shores, bogs; NC–Nfld. May–Aug.

Woody Plants

PINACEAE
Pine Family

Slash Pine
Pinus elliottii Engelm.

[422]
Pines are easily recognized by their evergreen needlelike leaves that are in clusters of 2–5, sheathed at their

bases by a number of scale leaves. The sheath drops off early in *P. strobus*. Male and female cones occur on the same tree. Few pines are shade-tolerant.

Slash pine is usually a rapidly growing tree, large at maturity. Needles are 17–25 mm long, in clusters of 2 and 3 on the same tree. Buds are covered with reddish scales edged with white to tawny. Seed cones are 6–15 cm long, on pedicels 20–25 mm long, and often reflexed. Abundantly planted. Common. Usually in moist soils but occasionally in dry soils—in a variety of habitats including stable dunes. MS–sSC. Jan.–Feb.

Loblolly and longleaf pines are often confused with slash pine. The former, *P. taeda* L., is distinguished by having needles in clusters of 3, or rarely of 3 and 4 on the same tree, and sessile cones. Common. In both dry and wet habitats—similar places; MS–sNJ. Mar.–Apr. Longleaf pine, *P. palustris* Mill., is distinguished by generally longer needles, 20–45 cm long, sessile cones, and silvery buds that are covered by prominently lacerated scales. The nongreen scales at base of leaf clusters are also prominently lacerated. Occasional. Well-drained sandy soils inland from dunes; MS–sNC. Mar.–Apr.

Shortleaf Pine
Pinus echinata Mill.

[423]
A large tree when mature. Plates of trunk bark with resin pockets. Needles 7–13 cm long, in clusters of 2 and 3 on same tree. Bark of 3-year-old twigs rough and flaking. Mature cones sessile, 3–7 cm long, opening at maturity, usually falling after 3 years. Common. Usually dry soils of a variety of habitats; MD–NY. Mar.–Apr.

P. clausa (Chapm. ex Engelm.) Vasey ex Sarg. (Sand Pine) is similar. Needles 5–12 cm long. Bark of 3-year-old twigs rough, tight, and not flaking. Cones persist for many years unopened or occasionally open at maturity. Common. Loose sands, includ-

ing dunes adjacent to beach; eAL–FL. Jan.–Feb.

Spruce Pine
Pinus glabra Walt.

[424]
Medium-sized tree at maturity. Bark of small trunks smooth and dark gray. On older trunks the bark is closely ridged and furrowed, suggestive of spruce bark. Needles in clusters of 2, 5–13 cm long, under 1 mm wide. Bark of 3-year-old twigs tight, nearly smooth, not flaking. Cones 5–10 cm long, sessile, with minute spines, if any. Quite shade-tolerant. Occasional. Usually occurs as scattered individuals in broadleaf woods of bottomlands and uplands; GA–sSC. Mar–Apr.

P. virginiana Mill. (Scrub Pine) is closely similar but needles are 2–8 cm long and cones have conspicuous slender spines 1–3 mm long. Occasional. Usually in poor soils in the open; VA–NY. Apr.–May.

Pitch Pine
Pinus rigida Mill.

[425]
A small tree at maturity. Needles 5–14 cm long, 1.5–2.0 wide, stiff, in clusters of 3, or rarely 3 and 4, usually falling their second year. Cones 3–7 cm long, about as broad as long, ovoid when closed, opening at maturity, then nearly globose, spines pyramidal and sharp-pointed. Common. Stable dunes, poor soils; nDE–ME. May.

P. serotina Dougl. (Pond Pine) is similar, but leaves 12–20 cm long, flexible, persisting 3–4 years. Cones usually remain closed for years. Occasional. Poorly drained pinelands that are frequently burned, pond margins; FL–sNJ. Apr.

P. strobus L. (White Pine) is easily recognized by the needles in clusters of 5, 8–12 cm long, slender, and pliable. Shade-tolerant. Rare. In woods or in the open; nNJ–Nfld. May.

CUPRESSACEAE
Cypress Family

Atlantic White-cedar
Chamaecyparis thyoides (L.) B.S.P.

[426]

Strong-scented evergreen tree, usually with a narrow cone-shaped crown. Clusters of small branchlets essentially in one plane, as if pressed. Leaves very small, opposite, sessile, scalelike. Seed cones at first green, glaucous, and globose; losing seeds in Oct.–Nov. and becoming brown and leathery, falling much later. Previous year's seed cones may be seen in the photograph, which was taken in June. The brown remains of the very small pollen-bearing cones may be seen on tips of some of the branchlets. Plants are shade-tolerant. Occasional. Depressions between and behind stable dunes, bogs; NJ–MA. Mar.–Apr.

Red-cedar
Juniperus virginiana L.

[427]

Small evergreen aromatic tree with a variable crown, mostly broadly pyramidal to narrowly columnar. Bark brown and shreddy. Leaves opposite and scalelike, as in the photograph; or flat, subulate, and in whorls of 3. Both types may occur on the same tree; seedlings almost exclusively have the latter type. Clusters of branchlets bushy, not appearing as if pressed. Seed cones mature the first year, are 5–7 mm long, berrylike, dropping without splitting open. Male cones, borne on separate trees, are 3–6 mm long. Common. Open places or thin woods of usually dry soils; VA–sME. Jan.–Apr.

J. silicicola (Small) Bailey (Southern Red-cedar) is quite similar, but the crown is broadly conical to rounded, the seed cones tending to be the smaller sizes. Probably best treated as a variety of *J. virginiana*. Common. Dry to wet places—dunes, swales, brackish flats, shell mounds, thin woods; eTX; MS–NC. Jan.–Feb.

J. communis L. var. *depressa* Pursh, a shrub, has linear sharp-pointed leaves 8–18 mm long and in whorls, seed cones 6–10 mm long and maturing the second year. Rare. Poor soils in open; nNJ–Nfld. Mar.–Apr.

ARECACEAE
Palm Family

Cabbage Palm
Sabal palmetto (Walt.) Lodd. ex Schult. & Schult.

[428]

A branchless tree to 25 m tall with a hemispheric crown of large evergreen fanlike leaves. Leaf segments filamentous on margins and usually at tip. Petioles not armed, tapering into the blade base, thus forming a midrib. Leaves are persistent after dying. Fruit a drupe 8–12 mm across. Common. In pure stands or mixed with other broadleaf trees or pines. Edges of ponds and marshes, including brackish ones, dunes; FL–sNC. June–July.

S. minor (Jacq.) Pers. (Dwarf Palmetto) is a shrub, has leaves with a midrib as in cabbage palm. The leaf segments are not filamentous. The stem is often solely underground, sometimes short ascending to erect, or in LA and TX to 6 m tall. Fruit a drupe 6–8 mm across. Common. Wet woods; TX–NC. May–July.

Saw Palmetto
Serenoa repens (Bartr.) Small

[429]

Easily recognized by its evergreen fanlike leaf blades and petioles armed with spines. Leaf blades green to blue-green, sometimes glaucous. Stems usually running along the ground hidden by leaf bases and litter but exposed after fire. Stems sometimes erect to 6 m tall, frequently with 1 or a few branches. Fruit a drupe 15–25 mm in diameter. Common. Dry habitats—pinelands, pine-deciduous woods, thin live oak or maritime woods, dunes; LA–sSC. Mar.–July.

AGAVACEAE
Agave Family

Mound-lily Yucca
Yucca gloriosa L.

[430]
Shrub or small tree to 5 m tall, only occasionally branching. Leaves linear, very numerous, firm but not rigid, to 60 cm long, sharp-pointed, margins smooth or sometimes lightly scabrous, and with a very narrow thin light-brownish line along the edge. Perianth segments 6, alike, white or nearly so, 4–5 cm long. Fruit fleshy to leathery, 5–6 cm long, drooping. Seeds flattened, 7–8 mm broad. Yuccas are sometimes put in the lily family. Common. Mostly in dune areas, including active ones; edge of brackish marshes; thin woods on sandy soils; MS; eFL–NC. Aug.–Nov.

Spanish-bayonet
Yucca aloifolia L.

[431]
Shrub or tree to 3.5 m tall, with a coarse stem, usually with 1 to a few branches. Leaves very numerous, linear, rigid, to 60 cm long, margins with small very sharp spiny serrations that can cut easily and deeply into the flesh of the unwary person. The leaves are also dangerously sharp-pointed. Flowers much like those of *Y. gloriosa* but a little larger, 5–6 cm. Berries, shown in the picture, are 7–9 cm long. Seeds are turgid, not flattened, 7–8 mm long. Common. Active and stable dune areas, edge of brackish marshes, shell mounds, and scattered in a variety of dry habitats in thin woods or in the open; LA–sNC. May–July.

Bear-grass
Yucca flaccida Haw.

[432]
Trunk very short to absent. Entire plant to 2.5 m tall, usually much shorter. Leaves linear, to 60 cm long, 15–60 mm wide, leathery, surfaces smooth or faintly scabrous, pliable, tapering to a sharp point but much less dangerous than the 2 preceding species, margins fraying into filamentous threads. Fruit an erect 3-carpelled capsule. Occasional. Dry places—thin woods or in the open, between stable dunes, old fields, roadsides; AL–MA. Apr.–Sept. *Y. filamentosa* L. in part; *Y. smalliana* Fern. as treated in most manuals.

Y. filamentosa L. is closely similar but the leaves are abruptly acute and slightly cusped at apex, and both surfaces of the leaves quite scabrous. Occasional. Similar places; nGA–VA. Apr.–Sept. *Y. filamentosa* in part as treated in manuals.

SMILICACEAE
Smilax Family

Dune Greenbrier
Smilax auriculata Walt.

[433]
A look at only leaves of any *Smilax* species can easily lead a person to think that they are dicots. Solid stems with scattered vascular strands and flower parts in 3's or multiples of 3 confirm that they are monocots. Woody *Smilax* species are easy to recognize. All are vines, climbing mostly by tendrils that arise in pairs from the leaf petioles. Prickles are often present on the stems. Fruits are 1- to 3-seeded berries. Identification to species is often difficult.

This species is prominent in seaside habitats. Branchlets are usually prominently zigzagged. Leaves glabrous, green on both sides, usually somewhat hastate, sometimes oblong, conspicuously veined beneath and usually above with a series of lateral veins that, at their extremities, are parallel to and within 0.5 mm of the margin. Petioles of the leaves that subtend the fruits are 8 mm long or less. Peduncles less than 1.5 times as long as their adjacent petioles. Berries often glaucous, black beneath when mature, maturing first season. Common. Dunes, dry sands of pinelands, maritime and live oak woods, fencerows; LA–NC. Apr.–July.

Fringed Greenbrier
Smilax bona-nox L.

[434]
Much like the preceding species but peduncles are 15–40 mm long, 1.5 or more times as long as the subtending petioles. Leaves are also different in having a continuous marginal rib (which is more distinct from the underside of the leaf and when the leaf is dried). Leaf margins completely smooth or with scattered fine spines. Common. Thin broadleaf woods, pine-hardwood woods, fencerows, shrub areas but rarely on dunes; TX–MD. Mar.–June.

Sawbrier
Smilax glauca Walt.

[435]
Easily recognized among the *Smilax* species by glabrous stems, and leaves that are glaucous beneath and have distinct veins in addition to the midvein. Blades ovate to lanceolate, margins entire, thin, and without a rib. Internodes of branchlets are not 4-angled or 4-ridged. Berries bluish black. Common. Dry to wet habitats—woods of various types, shrub areas, old fields, fencerows; AL–NY. Apr.–June.
S. rotundifolia L. has a similar leaf shape and also has a thin leaf margin but leaves are not glaucous beneath. Peduncles are 5–15 mm long and less than 1.5 times as long as the adjacent petioles. Branchlets are usually not zigzagged. Berries bluish black. Occasional. Stable dune areas, broadleaf woods, thickets, fencerows, old fields; NJ–sME. Apr.–June.

S. walteri Pursh has an aspect similar to the above 2 species but mature berries are red. Rare. Swamps, wet woods, pond margins; MS. Mar.–Apr.

Bamboo-vine
Smilax laurifolia L.

[436]
An extremely vigorous species with dead as well as live stems sometimes forming impenetrable entanglements. Leaves glabrous, evergreen, thick-coriaceous, oblong to oblong-linear or oblong-lanceolate to rarely broadly linear, lower part of the midvein more prominent than the laterals, the complete edge with a closely and evenly submarginal vein. Except under close inspection these veins appear marginal. Berries black; the only native species in which fruits ripen late in the second season after flowering the previous year. Common. Moist to wet places—pond margins, pinelands, swampy areas; MS–VA. July–Sept.

Sarsaparilla-vine
Smilax pumila Walt.

[437]
Plants trailing or low-climbing, to 50 cm long. Stipular tendrils sometimes absent on small plants but always present on larger ones. Stems and lower leaf surfaces densely hairy. Fruit bright red, a 1-seeded berry, ovoid, tapering to a blunt beak, ripening in the spring after flowering the previous year, then often persisting. Common. Sandy soils of maritime and live oak woods, pinelands; MS–SC. Sept.–Nov.

SALICACEAE
Willow Family

White or Silver Poplar
Populus alba L.

[438]
Populus species are trees with deciduous alternate simple petioled leaves. Flowers are in dense spikelike inflo-

rescences, male on one plant, female on another, appearing before or during leaf expansion. Fruits are small capsules. Seeds minute, with a tuft of long hairs.

White poplar is usually a small tree but to 30 m tall, often reproducing only by root suckers since only one sex is generally present at a given locality; the plants usually have been introduced as ornamentals. Leaf blades ovate, to 9 cm long, with a few coarse teeth. Plants are conspicuous, especially in the wind, because of the white undersides of leaves, this being due to dense felted hairs. The upper side may be glabrous or may be hairy, especially when young as in the picture. Occasional. Abandoned house sites, around buildings, fencerows; GA–NS. Mar.–May.

Populus alba

Large-toothed Aspen
Populus grandidentata Michx.

[439]
Tree to 18 m tall. Bark smooth, greenish white to cream-colored, becoming furrowed and gray to dark brown. Terminal bud ovoid, acute, the visable scales dull brown and finely hairy. Outer end of petioles flattened, leaf blades therefore easily flexed by the wind. Blades with 15–20 conspicuous teeth on a side. Occasional. Dry to wet habitats in the open or in woods; nMD–NS. Mar.–Apr.

P. tremuloides Michx. (Quaking As-

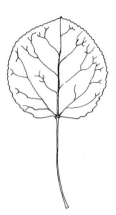

Populus tremuloides

pen) is the only other *Populus* species likely to be encountered in seaside habitats. Similar to the above species but with 20–40 smaller crenate-serrate teeth on a side of leaf blades, or nearly entire, and buds shiny and glabrous. Occasional. Swales, woods, cut-over woods; nNJ–Nfld. Mar.–Apr.

Swamp Willow
Salix caroliniana Michx.

[440]
Seaside willows are shrubs or trees with alternate lanceolate to ovate or oblong-oblanceolate leaves. Buds with only 1 scale. True terminal bud lacking, the actual stem tip dead. Twigs are tough and flexible but lateral ones are easily broken off at their base. Flowers are very small, without a perianth, in spikelike clusters, male on one plant, female on another. Fruits are small capsules. Seeds minute, with a tuft of long silky hairs. Bark bitter, containing tannic acid and salicin. The bark was used as an early source of medicine with qualities of aspirin. Of perhaps 100 species in North America only this species and *S. nigra* occur in seaside habitats south of NJ, and only the latter and 2 others are likely to be encountered from DE–MA.

Swamp willow is a shrub or small tree to 10 m tall with lanceolate to lance-ovate leaf blades that are light-green to glaucous beneath. Petioles to 2 cm long. Pedicels 1.5–4 mm long.

Fruits are glabrous. Common. Moist to wet freshwater habitats in the open, rarely in woods; FL–VA. Feb.–Apr.

S. nigra Marsh. (Black Willow), a shrub or tree to 20 m tall, has first-year twigs yellowish-brown to brown, leaf blades lanceolate to elongate-lanceolate and the same green color on both sides, petioles 2–8 mm long, and pedicels 1–2 mm long. Common. Moist to wet freshwater habitats in woods or in the open; TX–wFL; nGA–NBr. Feb.–Apr.

Long-beaked Willow
Salix bebbiana Sarg.

[441]
Shrub or small tree to 5 m tall. First-year twigs brown to dark brown. Leaves ovate to oblong or oblong-oblanceolate, glaucous and with a wrinkled surface beneath and grayish hairs on both sides. Fruits are hairy, long-beaked, the pedicel 3–6 mm long. Common. Moist to wet fresh-water places; nNJ–Nfld. Apr.–June.

S. discolor Muhl. (Pussy Willow) is a shrub or small tree to 7 m tall, also with leaf blades glaucous beneath, sometimes also sparsely hairy on underside only, and not wrinkled beneath. First-year twigs are brown to nearly black. Fruits are also hairy but pedicels only 1.5–3 mm long. Occasional. Freshwater habitats in woods or in the open; NY–Nfld. Apr.

MYRICACEAE
Wax-myrtle Family

Wax-myrtle
Myrica cerifera L.

[442]
Seaside *Myrica* species are aromatic rhizomatous shrubs or small trees with alternate simple leaves. Flowers are very small, without a perianth, unisexual, in small dense clusters, the male in elongate spikelike clusters, the female in globose clusters. The sexes almost always on separate plants. Fruits globose to ovoid.

This species is a shrub or tree to 10 m tall. Leaf tips acute, the sides forming an angle under 65°. Leaves ever-green, sometimes dropping in severe winters; the surfaces with abundant waxy granules, especially the under-side. Fruits 2.0–3.5 mm long, without hairs, with an irregular surface heavily coated with waxy granules. The wax of this and other species is removed by boiling in water. The wax was an important source of candle wax in the past. Common. Scattered in primary dunes, frequent in stable dune areas, edges of marshes and ponds, woods; LA–sNJ. Apr.–June.

Bayberry
Myrica pensylvanica Loisel.

[443]
Shrub or rarely a tree (to 4.5 m high) with year-old twigs light gray or drab. Leaves deciduous, thin, dull above, the margins at the tip on all or most leaves forming an angle over 65°. Young fruit densely hairy, mature fruit 3.5–5.5 mm across. Common. Habitats similar to those of *M. cerifera*; nNC–Nfld. Apr.–May.

M. heterophylla Raf. also has leaf tips forming an angle over 65°. It is distinguished by having blackish year-old twigs, leaves shiny above and coriaceous, sometimes ever-green, but deciduous in cold winters, young fruits glabrous, and mature fruits 3.0–4.5 mm in diameter. Rare. Similar places; MS–wFL. Apr.

Sweet-gale
Myrica gale L.

[444]
Rhizomatous shrub to 2 m tall, year-old branches reddish purple to brown or black. Leaves deciduous, oblanceolate, similar to those of *M. cerifera*, grayish, toothed towards the apex. Flowers and fruits are clustered near tip of previous year's leafless stem and above the point of attachment of leaf-bearing branches of the current year. In other deciduous seaside spe-

Myrica gale

cies, flowers and fruits are below the leaf-bearing stems of the current year. Occasional. Swamps, freshwater and brackish marshes, bogs, pond edges, swales; nNJ–Nfld. Apr.–May.

Sweet-fern
Comptonia peregrina (L.) Coult.

[445]
Highly fragrant rhizomatous shrub to 1.5 m tall, with hairy stems. Leaves deciduous, alternate, linear-lanceolate to linear-oblong, dark green above, pale beneath, hairy beneath and sometimes above also, conspicuously pinnately-lobed. Flowers small. Male flowers in spikelike clusters near tip of last year's stem (shown in photograph). Female flowers on separate plants, with 8 bracts persisting around their bases, the group of flowers developing into a burlike ball 1–2 cm thick. Fruit a nutlet 3–4 mm long. Occasional. Thin woods or in the open, dry usually sandy soils; NJ–NS. Apr.–May.

JUGLANDACEAE
Walnut Family

Pignut Hickory
Carya glabra (Mill.) Sweet

[446]
Hickories are trees with deciduous, alternate pinnately, once-compound leaves with 5–19 serrate leaflets with an acute to acuminate apex. Stems without prickles; pith small, continuous, soft; buds with overlapping scales. Bruised or crushed leaves with pungent odor. Flowers unisexual, the male ones small, with a calyx, but no corolla, in long spikelike clusters. Female flowers single or 2–6 in a spike. Fruit a hard nut surrounded by a firm 4-sectioned husk that usually splits apart at maturity. See photograph of *C. tomentosa*.

Pignut Hickory usually has 5 leaflets. They are glabrous and lack scales beneath at maturity. Bark shallowly furrowed, with low interlacing ridges forming a diamond pattern, not shaggy. Fruits pear-shaped, rarely globular, husk 2.5–3.5 mm thick, not splitting to base. Quite shade-tolerant. Common. In woods or in the open—stable dunes, mixed with most any other trees on well-drained soils; MS–MA. Apr.–May.

Mockernut Hickory
Carya tomentosa (Poir.) Nutt.

[447]
Medium-sized tree with tight bark that is shallowly ridged and furrowed into a diamond pattern. Leaves to 40 cm long. Petioles, rachises, and lower surfaces of the 5–9 leaflets with small clusters of hairs. Underside of leaflets with granular-resinous scales. First-year twigs stout, about 6 mm in diameter, bud scales overlapping. Fruit 3.5–5 cm long, husk 3–5 mm thick and slowly splitting to the base. Common. Dry woods, rarely in moist situations, near margins of salt marshes; nGA–MA. Apr.–May. Syn.: *Carya alba* (Mill.) K. Koch.

C. *ovata* (Mill.) K. Koch may be recognized by having shaggy bark; leaflets usually 5, glabrous beneath or with a few tufts of hairs on main veins, and serrations with small persistent tufts of hairs. Fruit globose or nearly so, 35–50 mm across, the husk 5–10 mm thick, splitting to base. Occasional. Usually moist places—mostly mixed with other trees; nNJ–sME. May.

Pecan
Carya illinoensis (Wang.) K. Koch

[448]
Pecan is easily recognized by having 11–19 lanceolate to ovate and usually falcate leaflets, bud scales not overlapping but valvate, fruit 4-winged but otherwise smooth, husk about 1 mm thick, splitting to base. Probably not native in seaside habitats but frequently escaped from plantings; the nuts are distributed principally by birds and rodents. Common. In thin woods or in the open, usually in moist habitats. AL–MA. Apr.–May.

BETULACEAE
Birch Family

Hop-hornbeam
Ostrya virginiana (Mill.) K. Koch

[449]
Small tree with flaking bark on trunk. Twigs with circular lenticels, a true terminal bud absent; the tip bud is the axillary bud of the uppermost leaf. Buds circular in cross section. Leaves simple, alternate, deciduous, 2-ranked. Blades with fine soft hairs beneath, symmetrical at base, margins sharply serrate to doubly-serrate. Fruits are nutlets, each enclosed by a membranous inflated sac, these arranged in a spike as seen in the photograph. Occasional. Well-drained soils—usually an understory tree, uncommonly in the open; nNC–VA; nNJ–NS. Apr.–May.
Leaves of *Carpinus caroliniana* Walt.

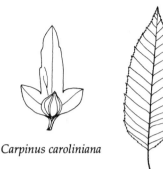

Carpinus caroliniana

Carpinus caroliniana

(Blue-beech or Ironwood) are similar to those of *Ostrya* but are glabrous below except for small tufts of hairs in vein axils. Bark and fruits are quite different. Blue-beech trunk is smooth and has musclelike ridges. Fruits are small ovoid ribbed nutlets, each at base of a conspicuous 3-lobed leaflike bract. Buds are somewhat 4-angled. Rare. Poorly drained soil in woods or in the open, rich woods; VA–sME. Mar.–Apr.

Hazelnut
Corylus americana Walt.

[450]
Deciduous shrub usually growing in clumps or large colonies, with slender round geniculate twigs. Buds sessile, ovoid, with 4–6 scales. Leaves ovate to nearly orbicular, glabrous or nearly so above, finely hairy beneath, margins finely serrate, tips acute to acu-

Corylus cornuta

minate, bases cordate. Petioles glandular-hairy. Twigs with soft continuous pith, no stipule scars, somewhat triangular leaf scars with 3 bundle scars, and lacking a true terminal bud. Male flowers in spikelike clusters formed during summer of previous year. Fruit a large bony-shelled nut enclosed within the base of 2 large leaf-like bracts. Rare. Thickets; nNJ–MA. Apr.

C. cornuta Marsh. (Beaked Hazelnut) is similar in most aspects but petioles not glandular hairy and leaf-like bracts around nut fused into a tubular beak 15–45 cm long. Rare. Thickets; NY–Nfld. Apr.

Paper Birch; Canoe Birch
Betula papyrifera Marsh.

[451, 451a]
Conspicuous tree with white bark that separates in thin layers. Leaves alternate, 2-ranked, deciduous. Blades ovate, tip acuminate, bases cuneate to rounded, margins sharply serrate, 3–8 lateral veins on each side, sparsely hairy beneath, often only on the veins. Male and female catkins as in photograph, the latter erect. Fruit in cylindrical conelike structures 2–6 cm long. Occasional. Dry to moist habitats—in woods or in the open, stable dunes, shores; NY–Nfld. Apr.–May.

B. populifolia Marsh. (Gray Birch) is a shrub or tree with similar leaves but glabrous beneath, doubly serrate, and long-acuminate. The bark does not separate by layers and is chalky-white with dark elongate markings. Occasional. Similar places; MD–NS. Apr.–May.

Betula populifolia

Betula papyrifera

River Birch
Betula nigra L.

[452]
Tree to 25 m tall. Bark of trunks that are 4–20 cm in diameter pinkish to tan or reddish brown, peeling in thin layers, becoming darker and scaly when larger. Leaves alternate, 2-ranked, deciduous, ovate to deltoid-ovate or ovate-oblong, acute, sharply doubly-serrate except at base of blade on each side of petiole where it is entire and parallel to lowest lateral vein. Blades light-colored and hairy beneath. Fruits erect cylindrical conelike structures to 3 cm long. Occasional. Swales, poorly drained soils in woods or in the open; MD–MA. Apr.

Smooth Alder
Alnus serrulata (Ait.) Willd.

[453]
Shrub or small tree to 10 m tall usually growing in clumps. Trunk smooth but longitudinally shallowly ridged and grooved, greenish-brown to grayish-brown. Lateral buds stalked, outer scales of equal length, scarcely overlapping. Leaves alternate, not 2-ranked, deciduous. Blades broadest at or above middle, margins finely and sharply serrate, principal leaves with 8–12 pairs of lateral veins. Flowers are unisexual, both sexes on same plant, borne in elongate conelike spikes. Male spikes conspicuous in spring before leaves appear. Fruits in dense ovoid conelike clusters to 18 mm long, dropping seeds in Aug.–Oct. of the same year, but seedless

"cones" persisting into following year, as in photograph. Common. Wet places, in woods or in the open; LA–FL; SC–sNC; nVA–NBr. Feb.–Apr.

A. rugosa (Du Roi) Spreng. (Speckled Alder) is quite similar but leaf blades are broadest at or below the middle and doubly-serrate with teeth of irregular sizes. Occasional. Similar places; MD–Nfld. Mar.–Apr. *A. maritima* Muhl. ex Nutt. has many features of the above 2 alders but the principal leaves have only 5–8 pairs of lateral veins, it flowers Aug.–Sept., and the fruits mature the following year. Rare. Wet places, in woods or thickets; MD–sDE. Mar.

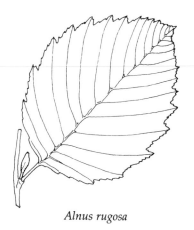

Alnus rugosa

FAGACEAE
Beech Family

American Beech
Fagus grandifolia Ehrh.

[454]
Tree to 35 m tall with thin smooth light gray bark without furrows and ridges. Easily recognized by leaves that have evenly spaced lateral veins, each ending at tip of a marginal tooth; and by stipules that soon fall off leaving scars that encircle the twig. Buds are also distinctive in that they are long (to 25 mm), slender, lance-shaped, and sharp-pointed.

Fruits are 3-sided sharply angled nuts enclosed in pairs by 4 leathery to woody bracts covered with soft bristlelike spines. Occasional. Rich woods, depressions between old stable dunes; VA–NS. Mar.–Apr.

Chinquapin
Castanea pumila (L.) Mill. var. *ashei* Sudw.

[455]
Shrub or small tree, sometimes suckering from roots. Bark rough. Mature terminal bud of twigs ovoid, finely hairy, about 3 mm long. Vigorous twigs with a single leaf at tip. Leaves densely fine-hairy to nearly glabrous beneath. Lateral veins on each side evenly spaced, parallel, each terminating in a sharp tooth of the widely serrated margin. Flowers are unisexual, both sexes borne on same plant. Male flowers crowded on long erect to ascending spikelike structures. Fruit an acornlike nut surrounded by a bur with stiff sharp spines, the bur splitting into 2–4 parts at maturity. Burs may be found persisting on the ground beneath the shrubs or trees. Occasional. Thin woods or in the open in sandy soils on stable dunes, behind dunes, or further inland; MS–sVA. June–July.

Water Oak
Quercus nigra L.

[456]
Oaks are shrubs or trees with alternate leaves, 2–4 of them clustered at each twig tip. This should be checked on vigorous twigs because species without 2 or more leaves at tip of vigorous twigs often have leaves clustered on stubby twigs. Fruits are acorns, a type of nut, borne in scaly cups. Any part of an oak tree may be poisonous when eaten in considerable amounts because of the tannic acid content. Oaks may be divided into black and white oak groups. Black oaks have bristle tips on apex and any lobes of leaves (see photograph of *Q. coccinea*), and acorns that mature after 2 growing seasons. Thus

small acorns are on the trees during winter. White oaks have no bristle tips on leaves or their lobes, and acorns mature and fall the first year. We have recorded 21 species of oaks in seaside habitats. Of these, 3 are unlikely to be encountered and are omitted.

Water oak is a rapid-growing tree to 35 m tall. It usually loses leaf bristles early. Acorns on the previous year's twigs, as in the photograph, or immature acorns on the current year's twigs during fall and winter mark it as a black oak. Leaves are prominently broadest at the tip, tapered at the base, only 5–10 cm long, rarely to 15 cm, mostly deciduous but some usually persisting into winter. Common. Moist to dry soils in woods or in the open—wooded dunes and inland; eTX; MS–DE. Apr.

Blackjack Oak
Quercus marilandica Muenchh.

[457]
This tree's leaves lose their bristles early but the tree can be identified as a black oak by mature acorns on the previous year's twigs. Also no seaside white oak has leaves as large and of this general shape. Blackjack oak is a small tree, rarely over 15 m tall, with deciduous leaves, drooping lower branches, and rounded crown. Leaves broadly obovate to triangular-obovate, 7–25 cm long, 7–20 cm wide, with 3–5 short lobes at apex, or rarely unlobed. Winter buds slender, brown, to 1 cm long, with rusty hairs. Common. Thin pinelands or mixed woods, or in the open; on well-drained soils; MS–sMA. Apr.

Scarlet Oak
Quercus coccinea Muenchh.

[458]
Tree to 25 m tall with deciduous, deeply lobed leaves that are green on both sides and quite shiny above. Except when leaves are young the only hairs are small tufts in axils of major lateral veins on undersides. In the photograph, note bristle tips on

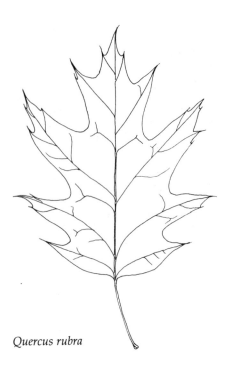

Quercus rubra

leaves and 2 age classes of acorns, the younger ones in the axils of the third and fourth leaves from the bottom. Upper scales of acorn cup loose at tips. Acorn often with fine concentric grooves near the apex. Occasional. Woods between stable dunes, dry well-drained soils; MD–sME. Apr.

Q. rubra L. (Northern Red Oak) is similar but leaves dull green above, lobed usually only halfway to midrib or less (measure perpendicular to midvein) and tip of scales of acorn cup are tightly appressed. Common. Stable interdunes, well-drained sites; nVA–NS. Apr.–May.

Black Oak
Quercus velutina Lam.

[459]
A tree to 20 m tall with deciduous leaves, these shallowly lobed, like those of northern red oak, to deeply 5- to 9-lobed, like those of scarlet oak, but the midrib of the upper surface with small inconspicuous hairs. These hairs can often be detected by gently drawing finger tip along the midvein.

Leaves are rarely hairy on entire undersurface. Apex and lobes of leaves are bristle-tipped. Scales of the acorn cups hairy on back and have free tips forming a loose fringe at top of cup. Common. Dry sites, stable dunes and inland; VA–sME. Apr.–May.

Southern Red Oak; Spanish Oak
Quercus falcata Michx.

[460]
Tree to 40 m tall with deciduous leaves. Easy to recognize by the bell-shaped base on the leaf blades and the pale undersides due to small dense soft gray to light rusty gray hairs. Individual hairs are not easily seen and perhaps cannot be detected at all late in the growing season. Main lobes of the leaf are bristle-tipped and usually falcate. Terminal buds ovoid, pointed, reddish brown, to 8 mm long, the backs of the scales with brownish hairs. Scales of acorn cup tightly appressed, with dense gray hairs on back, the margins reddish. Common. Dry places—stable dunes and inland; MS–SC; VA–NY. Apr.–May.

Turkey Oak
Quercus laevis Walt.

[461]
A small tree, rarely tall, to 20 m. Leaves often persisting after dying in fall. Petioles usually twisted so that blades are vertical to ground. Blades coriaceous, mostly deeply cleft, with the 3–7 lobes separated by broad open sinuses, the 2 largest lateral lobes usually narrow and falcate; when mature, glabrous on both sides except for tufts of hairs in axils of the main lateral veins; bases acute. Acorn cups are unusual in being top-shaped and with the upper scales turned about ⅓ of way down inside cup. Occasional. Well-drained sandy soil, open pinelands and scrub areas; AL–sNC. Mar.–Apr.

Bear Oak
Quercus ilicifolia Wang.

[462]
Shrub or small tree to 6 m tall with deciduous leaves. Easily recognized by the character of the leaves and having first-year twigs densely covered with small gray felty hairs that gradually become thinner, the twigs nearly black the next year. Mature leaf blades are ovate to elliptical or obovate, 5–12 cm long, the 3–7 short triangular to ovate lobes with bristle tips, dark green above, gray felted hairy below. Photograph shows young leaves with hairs on both surfaces (as is true of many species), the previous year's twig still hairy, immature year-old acorns, male flowers in the pendent spikes, and a female flower in a leaf axil. The sinuses in the leaf margins become much more shallow with growth. Common. Well-drained sandy or rocky soils, scrub areas, thin pine woods; NJ–ME. May.

Quercus ilicifolia

Myrtle Oak
Quercus myrtifolia Willd.

[463]
Shrub to small scrubby tree to 9 m tall with thick evergreen leaves. One of the black oaks that loses bristle tips on leaves early. Rarely, acorns mature the first year. First-year twigs with a few to many tawny stellate hairs, sometimes nearly hiding the surface, sometimes becoming nearly glabrous but dull-colored. Leaves widest towards the tip or uncommonly near

the middle, ovate to elliptic or most commonly obovate, margin occasionally undulate, glabrous beneath at maturity except for small pedicellate stellate hairs in axils of main veins. Cup covering ¼ to ⅓ of acorn. Easily confused with *Q. chapmanii*, a white oak, but the latter has deciduous leaves, finely grayish hairy first-year twigs, acorns always maturing the first year, and cup covering about ½ of acorn. Common. Sandy pinelands, scrub areas, dunes; MS–sSC. Feb.–Mar.

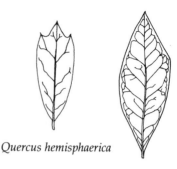

Quercus hemisphaerica

Quercus hemisphaerica

Diamond-leaf Oak
Quercus laurifolia Michx.

[464]
A tree to 20 m tall. A black oak but bristle tips on leaves dropping early. First-year twigs glabrous. Leaves generally deciduous, some often persisting through the winter. Most leaves widest at or near the middle, rarely some towards tip. Blades with tufts of hairs in axils of main veins below, surface generally dull, especially above, veins on upper side not prominent. Tips of most leaves obtuse, always some rounded, rarely a few acute. Occasional. Moist to wet soils in broadleaf woods; LA–MD. Mar.–Apr.

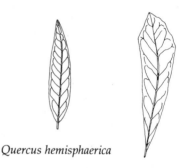

Quercus hemisphaerica

Quercus hemisphaerica

   Q. hemisphaerica Bartr. (Laurel Oak) has a similar aspect and is often confused with *Q. laurifolia*. Leaves are mostly evergreen, usually dropping in early spring. Laurel oak is recognized by having leaves quite glabrous beneath, surfaces shiny, tips mostly acute, rarely a few obtuse, veins on upper side very prominent. An occasional plant, especially young ones, may have a few to many leaves with 1–4 short acute-tipped lobes. Common. Stable dunes, maritime woods, thin pinelands, live oak woods; MS–NC. Feb.–Apr.

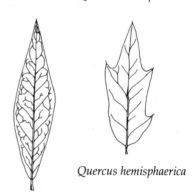

Quercus hemisphaerica

Quercus hemisphaerica

Bluejack Oak
Quercus incana Bartr.

[465]
Shrub or tree to 7 m tall. A black oak with leaves similar in shape to some

of those of diamond-leaf and laurel oaks and easily confused with them. Acorns are similar also. Leaves are generally deciduous but may persist during mild winters. The species may be recognized by leaves that are shiny above, the lower surface with many grayish stellate hairs, the rays of these hairs not parallel with the leaf surface but above a 40° angle with the leaf surface. By late fall or in winter most of these hairs may have fallen. Look for them beside the midvein near the base of the blade. Leaves may rarely have 1–4 short bristle-tipped lobes on leaves of second growth of year or on root or stump sprouts. Occasional. Dry sandy soils of scrub oak and thin pineland habitats; MS–sVA. Apr. Syn.: *Q. cinerea* Michx.

White Oak
Quercus alba L.

[466]
Tree to 30 m tall with deciduous leaves. For identification purposes, note absence of bristle tips on leaf lobes and maturing acorn on current year's twig, both shown in the photograph. First-year twigs glabrous, buds globose to broadly ovoid. Leaves 10–20 cm long; with 7–11 uneven, rounded lobes, the sinuses extending evenly and usually ⅓ or more to the midvein; broadest at or near the middle; glabrous and light green beneath but not pale. Scales of acorn cup tightly appressed, the basal ones much thickened. Occasional. Broadleaf woods, well-drained soils; nGA–sME. Apr.

Post Oak
Quercus stellata Wang.

[467]
One of the white oaks, a tree to 20 m tall with deciduous leaves. The broad cross made by the 2 largest lobes of the leaf has been given much emphasis in identification of this species but some leaves on most trees and occasionally all leaves are not "cross-shaped." Two other characters must be checked: (1) at least the upper ⅕ of the first-year twigs is densely clothed with fine grayish or tawny hairs that become matted until individual hairs are difficult to see even under magnification and the shiny cuticle is not visible; (2) underside of leaves has numerous sessile stellate hairs with rays horizontally spreading. These hairs gradually disappear but the

underside remains dull from their remains. Not shade tolerant. Common. Dry, often poor, soils—thin pinelands, pine-palmetto woods; thin live oak woods; MS–sMA. Apr.–May.

Q. margaretta Ashe (Shrubby Post Oak) has similar-shaped leaves but they are smaller, 4–10 cm long compared to 8–15 cm, and the underside has abundant to quite scattered pedicellate stellate hairs with the cuticle always easily seen. The shiny cuticle on the twigs is also readily visible; the twigs are glabrous to beset with numerous pedicellate hairs. Rare. Similar places; AL; nSC–sNC. Apr.

Chapman Oak
Quercus chapmanii Sarg.

[468]
A white oak. Tree to 9 m, leaves dying in fall but some may persist until late spring. First-year twigs closely and finely grayish hairy, the cuticle rarely visible under the hairs. Leaves 4–9 cm long, mostly obovate, rarely nearly elliptic, glabrous or with small scattered hairs beneath, margin even or wavy or shallowly lobed near tip. Easily confused with *Q. myrtifolia*, which see for separating characters. Occasional. Scrub oak pinelands, stable dunes; MS–sSC. Feb.–Mar.

Sand Live Oak
Quercus geminata Small

[469]
A white oak. Shrub to medium-sized tree. First-year twigs with very small dense hairs, sometimes becoming nearly glabrous late in year. Leaves coriaceous, evergreen, commonly entire and unlobed but on sprouts and scrubby plants often with rounded lobes and/or sharp coarse teeth. Leaf margins are rolled and hiding the very edge of the underside. Upper surface with depressed midvein and lateral veins. Undersurface often appearing glabrous but always with very small sessile horizontally radiating hairs as in *Q. virginiana*, but unlike the latter in also having raised hairs, which may be quite scanty. Im-

portant component of dune habitats. Common. Dunes, leached sands elsewhere; MS–sNC. Mar.–Apr. Syn.: *Q. maritima* (Michx.) Willd.

Live Oak
Quercus virginiana Mill.

[470]
Much like sand live oak. Ranges in size from shrubby plants to individuals with trunks over 1 m in diameter and rounded crowns with spreads of over 30 m. In dense woods the crown may be high. Live oak may be distinguished from *Q. geminata* by unrolled leaf margins, or, if rolled, none of the undersurface hidden from view by the rolled margin when observed with eye perpendicular to lower surface. Also veins are little sunken above and there are no raised hairs on the underside. Usually absent from dune areas. Common. Dry sandy soils—in pure stands or mixed with other species, including pines and palms; in the open; TX–VA. Mar.–Apr.

ULMACEAE
Elm Family

American Elm
Ulmus americana L.

[471]
Deciduous tree to 30 m tall with simple, alternate, 2-ranked leaves. Leaves 5–15 cm long, widest part of blade about midway or beyond, tip acuminate, base asymmetrical, margins doubly-serrate with more teeth than main lateral veins, axils of main lateral veins on underside with small tufts of hairs that diminish with age, veins on each side of midvein evenly spaced and parallel to each other. Hairs on twigs and buds absent or small and sparse. Twigs with circular lenticels, true terminal bud absent. Leaf characters of some of the other species overlap, making positive identification sometimes dependent

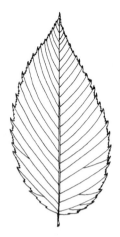

Ulmus americana

on characters of the fruits, which appear before leaves are developed. Fruits are 10–13 mm long, winged, the margins prominently ciliate, and the faces glabrous. Common. Wet woods, rich upland woods, occasionally in the open; eAL–Nfld. Feb.–Apr. Syn.: *U. floridana* Chapm.

U. alata Michx. (Winged Elm) is similar but often has cork wings on twigs, no tufts of hairs in axils of lateral veins of the leaf blades, and fruits with hairy faces. Rare. Dry soils, in woods or in the open; wFL; SC–sVA. Feb.–Mar.

Sugarberry; Hackberry
Celtis laevigata Willd.

[472]
Rapid-growing deciduous tree to 30 m tall with simple, alternate, 2-ranked leaves. Three major veins arise from the commonly asymmetrical base of the blade. Blades broadly lance-shaped, usually more than twice as long as wide, long acuminate at apex, margins entire or rarely with a few small teeth. Pith of mature twigs with closely and usually irregularly spaced chambers separated by soft irregularly sized partitions, sometimes nearly continuously hollow. Fruit a drupe 4–8 mm long, with a thin sweetish flesh, at maturity orange to brown or reddish. Common. Wet to dry places—stable dune

areas, pond margins, live oak and maritime woods, mixed hardwoods, in the open; LA–VA. Apr.–May.

C. occidentalis L. is similar in many aspects, differing mainly in having leaf blades broadly ovate, usually less than twice as long as wide, and margins serrate, at least on the upper half. Fruit 7–11 mm long, dark orange-red to purplish black. Rare. Moist to dry places, in woods or in the open; VA–MA. Apr.–May.

Celtis occidentalis

MORACEAE
Mulberry Family

Red Mulberry
Morus rubra L.

[473]
Tree to 15 m tall with simple, alternate, palmately veined leaves. Leaf margins simply serrate, occasionally with 1 or a few lobes, upper surface faintly to strongly roughened. Main basal veins 3. Fresh twigs with cloudy or milky juice, pith continuous. True terminal bud missing. Unlobed leaves much like those of basswood. Mulberries separated by having mature buds symmetrical at base, with 5–6 exposed scales, the first scale about ¼ as long as bud. Fruit fleshy, dark purple, cylindrical, 15–30 mm long, of many compact segments, each developed from a separate flower. Common. Plants usually scattered and in a variety of moist to dry habitats—live

and maritime oak woods, mixed hardwoods, shell mounds, fencerows, around buildings; MS–sMA. Apr.–May.

M. alba L. (White Mulberry) is similar but leaves are smooth above and more frequently and abundantly lobed. Fruit 10–20 mm long, white to reddish or light purple. Hybridized with *M. rubra* and having numerous intermediates. Occasional. Around buildings, fencerows, fields; eTX–sMA. Mar.–May.

LORANTHACEAE
Mistletoe Family

Mistletoe
Phoradendron serotinum (Raf.) M. C. Johnst.

[474]
Partially parasitic on trees and large shrubs. Leaves evergreen, thick, leathery, opposite, simple, entire, orbicular to elliptic or oblanceolate, to 13 cm long. Fruit a white to slightly yellowish berry, globose to ellipsoid, 2- to 3-seeded, 4–6 mm across. Flowers unisexual on separate plants. Common. On a variety of broadleaf trees; MS–sNJ. Sept.–Oct.

CHENOPODIACEAE
Goosefoot Family

Perennial Glasswort
Sarcocornia perennis (Mill.) A. J. Scott

[475]
A trailing or weakly arching to erect shrub lacking foliage leaves, these represented by small opposite appressed scales. Stems succulent. Flowers deeply sunk in scale axils. In the photograph stamens are seen protruding from leaf axils of the current year's stem. The previous year's stem is tan. Other glassworts are erect annuals of the genus *Salicornia*. Common. Salt and brackish marshes, sea

beaches, overwash areas, tidal flats; TX–NH. July–Oct. *Salicornia virginica* as used by various manuals but not *S. virginica* of Linnaeus.

BATACEAE
Saltwort Family

Saltwort; Vidrillos
Batis maritima L.

[476]
Trailing to arching or erect plants, often forming dense masses. Young stems and leaves succulent, crunching underfoot similarly to glassworts. Old stems woody with a pale brown bark that is very soft and easily sheds irregular flakes. Leaves and young stems are pale-green to yellow-green. Leaves sessile, curved, half-terete to nearly terete. Flowers clustered, unisexual, sexes on separate plants. Fruit an aggregate from 4–12 flowers, fleshy, 7–12 mm across, purplish black at maturity. Common. Salt marshes, salt flats, brackish marshes, muddy seashores, drift zones; TX–sSC. May–July.

MAGNOLIACEAE
Magnolia Family

Southern Magnolia; Bull Bay
Magnolia grandiflora L.

[477]
In all magnolias the bud is covered by 2 stipules that belong to the embryonic leaf inside the bud. The stipules fall early, leaving distinctive scars encircling the twig. Flowers with many separate spirally arranged stamens and pistils. Fruit conelike, each segment splitting open and releasing 1 or 2 red berrylike seeds that hang for some time on long extensible threads.

   M. grandiflora is a prominent tree with shiny coriaceous evergreen leaves 10–30 cm long. Sepals 3, petals 6–12 and 5–10 cm long. Common.

Dunes, scrub areas, hammocks, swamp woods, maritime woods, live oak woods; MS–sNC. May–June.

Sweet Bay
Magnolia virginiana L.

[478]
Shrub or tree to 18 m tall. Leaves 8–15 cm long; persisting into winter, a few remaining until spring in the south; undersides glaucous, quite noticeable even at a distance when the wind blows. The 3 sepals are easily seen in the photograph. Petals usually 8 but vary to 12, only 25–50 mm long. Occasional. Swales between stable wooded dunes; depressions in live oak, pine, and hardwood woods; pond margins; MS–NY. Apr.–July.

ANNONACEAE
Custard-apple Family

Papaw
Asimina parviflora (Michx.) Dunal

[479]
Shrub or rarely a small tree to 6 m tall. Crushed or bruised leaves with a strong odor offensive to most people. First-year twigs with reddish hairs. Terminal bud long, reddish-hairy, without scales. Leaves alternate, 2-ranked, margins entire, to 18 cm long and 10 cm wide. Blades oblong-obovate, base symmetrical, tips acute to short acuminate. Flowers 12-18 mm broad, with 3 somewhat triangular sepals and 2 series of 3 petals each. Stamens numerous, in a compact ball. Pistils 3–7. Four very young fruits, developing from a single flower, may be seen in the photograph. Mature fruits are 3–6 cm long, oblong-cylindric to rarely nearly globose, soft, aromatic, with a few large shiny brown seeds. Occasional. Scattered as understory plants in live oak woods, mixed hardwoods; AL–sNC. Feb.–Apr.

LAURACEAE
Laurel Family

Camphor-tree
Cinnamomum camphora (L.) Presl

[480]
Evergreen tree to 12 m tall, reproducing by root suckers but more widely from seeds. Foliage and twigs notably aromatic when bruised or crushed. Leaves simple, alternate, entire, with yellowish calluslike growths in principal vein angles on the upper side. Fruits are globose drupes about 1 cm across, seated on a pedicel shaped somewhat like a golf tee. Occasional. Fencerows, around buildings; scattered in most any type of woods except wet. TX–sSC. Apr.–May.

Redbay
Persea borbonia (L.) Spreng. var. *borbonia*

[481]
Aromatic shrub or small tree to 15 m tall. Leaves alternate, evergreen, undersurface light-colored and with sparse very small copper-colored to tawny appressed hairs, these invisible to the unaided eye. Fruit a drupe, 7–12 mm long. Leaves sometimes confused with those of *Magnolia virginiana* but absence of stipule scars on redbay will separate the two. Fruits are drupes 12–15 mm long, with 2 whorls of 3 each, persistent, enlarged perianth parts. Common. Dry habitats, dunes, scrub, maritime forests, hammocks, peaty soils; AL–sNC. May–June.

P. palustris (Raf.) Sarg. (Swampbay) is often confused with *P. borbonia* but undersurface of the leaves has erect hairs, often detectable by feel, usually can be seen by the unaided eye, but sometimes quite sparse. Common. Poorly drained soils in woods or in the open, pond margins, rarely in dry places; LA–VA. May–June.

Sassafras
Sassafras albidum (Nutt.) Nees

[482]
This shrub or small tree is easily recognized by its spicy aromatic odor and its 4 leaf forms, 1 unlobed and 3 lobed. Often called "Mitten tree" because some leaves have a single "thumb" on the left side, some on the right, and others on both sides. A few plants have only unlobed leaves but most have all 4 forms. Usually unnoticed are the flowers, which are male and female, the sexes on different plants. Fruit seated on a pedicel shaped somewhat like a golf tee, dark blue, fleshy, with a single seed. The bark of the root is used for tea and is the source of flavoring in root beer. Young leaves and tender stems are dried and ground into a powder, called filé, that is used in soups and gumbos. Common. Well-drained places—behind stable dunes, maritime forests, fencerows, margins of woods, shrub zone; MS–sME. Mar.–June.

SAXIFRAGACEAE
Saxifrage Family

Wild Gooseberry
Ribes hirtellum Michx.

[483]
Shrub with alternate, deciduous, palmately veined and lobed leaves. Stems usually not prickly. Flowers 1–3, on a short peduncle, not racemose. Petals 5, equal. Stamens 5, longer than petals, about as long as sepals. Ovary inferior. Fruit a reddish to purplish black, many-seeded, glabrous berry about 11 mm long. Occasional. Moist woods, stable dunes, thickets, swales, depressions; NY–Nfld. Apr.–June.

HAMAMELIDACEAE
Witch-hazel Family

ROSACEAE
Rose Family

Sweetgum
Liquidambar styraciflua L.

[484]
Tree to 30 m tall with alternate, simple, deciduous, aromatic, palmately veined and lobed leaves. Major veins and lobes 5, occasionally with 2 more smaller lobes at base. Entire blade starlike in appearance. Sometimes confused with maples, which can be separated readily by their opposite leaves. Twigs often have corky outgrowths that take the form of wings, ridges, or warts. Flowers are unisexual, the sexes on the same tree. The female flower consists of an ovary with 2 long styles and has 2 small scales at the base. These flowers are borne in large numbers in a tight spherical head at end of a long drooping stalk. These heads mature into tough drooping globose balls 25–30 mm across. They contain a few seeds and many aborted ones that resemble sawdust. Common. Poorly drained or rich woods of various types, spreading by winged seeds into nearby open areas; LA–NY. Apr.–May.

Witch-hazel
Hamamelis virginiana L.

[485]
Shrub or tree to 9 m tall with simple, alternate, deciduous 2-ranked leaves. Leaf margins entire and wavy to coarsely crenate or coarsely crenate-serrate, the base asymmetrical. Lateral veins 5–7 on each side, nearly straight, extending to blade margin. Buds hairy, without scales, the lateral ones stalked. Flowers in short axillary clusters. Petals linear, yellow. Fruit a hard, hairy, half-inferior, ovoid, 2-beaked capsule containing 2 shiny black seeds. Occasional. Rich, moist, or, uncommonly, dry woods; spotted along coast MS–NS. Oct.–Dec.

Meadow-sweet; Spiraea
Spiraea latifolia (Ait.) Borkh.

[486]
Shrub to 1.5 m tall with simple, alternate, glabrous, deciduous leaves. Inflorescence a panicle, mostly pyramidal, the branches glabrous. Sepals spreading. Petals white to pink. Stamens many, on rim of the cup-shaped hypanthium. Ovaries and fruits mostly 5 from each flower, glabrous, fastened to the inside bottom of the hypanthium, their sides free. Individual fruits are firm follicles. Common. Wet to dry places in thin woods or in the open—bogs, swamps, heath, swales, brackish marsh borders; NY–Nfld. June–Aug.

S. tomentosa L. (Hardhack) is similar but leaves are felty hairy beneath, mature sepals reflexed, and ovaries and fruits hairy. Common. Unusual in dry places but otherwise similar places; NY–NS. July–Aug.

Red Chokeberry
Aronia arbutifolia (L.) Pers.

[487, 487a]
Aronia species are deciduous shrubs to 3 m tall, reproducing from root sprouts as well as seeds. Leaves simple, alternate. Flowers in corymbs. Sepals and petals, 5. Carpels 5, united at base and fused with the surrounding hypanthium. Fruit a small pome 6–10 mm long. Sometimes included under *Pyrus* or *Sorbus*.

In red chokeberry the leaves are densely gray-hairy beneath and usually turn scarlet in the fall. The bright red fruits persist into winter. Common. Moist to wet places—thin hardwoods, depressions, thin pinelands, swales, scrub areas; eTX–NS. Mar.–May.

A. melanocarpa (Michx.) Ell. (Black Chokeberry) is similar but leaves are glabrous and do not turn red in the fall. Fruits are black and ripen and drop earlier. Intermediates with red chokeberry are common in many areas. Common. Similar places and also better-drained sites; CT–Nfld. Apr.–June. Syn.: *A. prunifolia* (Marsh.) Rehd.; *Pyrus floribunda* Lindl.

Wild Crabapple
Malus angustifolia (Ait.) Michx.

[488]
Shrub or small tree to 9 m tall. Leaves alternate, simple, deciduous, not 2-ranked. Stipules present, attached to lower end of petiole. True terminal bud present. Flowers in simple umbels. Petals nearly rose to pink, sometimes fading in age to white. Very fragrant when in flower. Fruit a green pome 25–35 mm long. Occasional. Thin woods, thickets, shrub zone in stable dune areas; SC–sNC; MD–sNJ. Apr.–May.

Serviceberry; Shadbush
Amelanchier arborea (Michx. f.) Fern.

[489]
Amelanchier species are deciduous shrubs or small trees without thorns. Buds are sessile, a true terminal bud present. Lenticels circular. Leaves are simple, alternate, 2-ranked, the margins serrate. Stipules are on lower part of petiole and fall early (they may be seen in photograph of *A. laevis*). Sepals and petals, 5. Stamens many. Ovary inferior. Fruits are pomes 6–10 mm across with many small seeds. Descriptions of the species in various manuals and in studies of the genus disagree in many respects and no common interpretation is available. Identification is also made difficult by numerous hybrids. There seem to be 3 "kinds" of serviceberries in seaside habitats. We describe them under the most appropriate names.

A. *arborea* is a shrub to small tree to 12 m tall. Leaves are hairy beneath at maturity, sometimes only sparsely so. Petioles hairy at maturity. Flowers several to many in declined or drooping racemes. Petals narrowly oblong to narrowly cuneate, 10–15 mm long.

Fruits reddish purple, insipid. Common. Dry areas—thin woods, thickets, stable dunes; VA–Nfld. Mar.–May.

Smooth Serviceberry
Amelanchier laevis Wieg.

[490]
Tall erect shrub or small tree to 15 m tall. Quite similar to *A. arborea* but leaves are glabrous beneath at maturity and glabrous or nearly so when plant is in flower. Fruit blackish purple and sweet. Common. Stable dune areas, thickets, thin woods; NY–Nfld. Mar.–May.

A. *canadensis* (L.) Medic. is a bushy shrub to a small tree, often in clumps, the racemes erect, petals oval to elliptic and 3–10 mm long. Fruit red to purple, sweet. Leaves densely hairy when young, becoming nearly glabrous late in year. Common. Low places—stable dune areas, edge of marshes, thin woods; AL; SC–Nfld. Mar.–May.

Hawthorn; Haw
Crataegus brainerdii Sarg.

[491]
Crataegus species are shrubs or small trees with simple alternate leaves and usually crooked thorny branches. Ovary inferior. Fruit a small pome with 1–5 bony nutlets inside, each containing a seed. They flower in early spring. A very large and complex genus, growing in a great variety of habitats, being most abundant in eastern North America. Identification of plants is often difficult, sometimes impossible. Most vegetative studies of seaside habitats report only the presence of the genus and do not indicate the species involved.

This species is similar to several occurring in seaside habitats. It is quite variable; at least 11 varieties have been named. The photograph is of a plant hanging over a brackish graminoid marsh. It is useless here to try and give distinguishing characteristics. In seaside habitats haws are to be expected mostly north of VA, occurring into NS. Mar.–May.

Dewberry
Rubus trivialis Michx.

[492]
Shrub with weak trailing biennial stems that usually root at tips and that bear numerous short prickles, these mostly glandular, slender, and small-based. Leaves compound, somewhat evergreen, glabrous or nearly so beneath. Leaflets 3 on flowering stems, 5 on first-year stems. Flowers usually 1 on each branch. Petals longer than sepals. Fruit black, composed of compact clusters of small drupes, the whole separating from the pedicel with the central receptacle included. Common. Thin dry woods, dunes, drift zone of low-energy shores, fencerows, roadsides, old fields, pastures, salt-spray areas; eTX–VA. Mar.–June.

R. *hispidus* L. is similar but the stiff bristles on the stem are not glandular, leaflets always 3, and there is more than 1 flower per inflorescence. Common. Dry to moist places, thin woods or in the open, rocky shores, thickets, roadsides; VA–NS. May–Aug. Another seaside dewberry is R. *flagellaris* Willd., which has coarse, often hooked prickles with expanded bases. Flowers 2–5, rarely 1, per inflorescence. Leaflets 3–5 on first-year stems, 3 on second-year. Common. Similar places; LA; VA–MA. May–June.

Blackberry
Rubus cuneifolius Pursh

[493]
Blackberry species have erect to arching biennial stems with prickles that are usually hooked. Leaves also with hooked prickles. Seaside species have leaves with 5 leaflets on first-year stems and 3 on second-year stems. Fruits like those of dewberries. Identification to species is difficult, there is little agreement among manuals concerning many species, and there is a large discrepancy regarding the number of species recognized. However, 3 kinds of blackberries are common in seaside habitats, can be recognized fairly easily, and can be designated fairly clearly by the names used here.

R. *cuneifolius* grows to 1.6 m tall. Leaflets oblanceolate to obovate, broadest above the middle, apex acute to rounded, whitish and softly hairy to the touch underneath. Common. Dry places—fencerows, old fields, pastures, thin pinelands, thin live oak woods; eFL–NY. May–June.

R. *argutus* Link is similar but leaflets elliptic to elliptic-lanceolate, the apex acuminate, broadest at or above middle, the soft hairs on the underside of the leaflets not grayish or whitish, the underside light green instead. Common. Similar places; eTX–NJ. Apr.–May. In the third species, R. *betulifolius* Small, leaflets are glabrous or nearly so beneath and broadest at or above middle. Common. Similar places, but also in moist habitats; AL–sNC. Apr.–May.

Red Raspberry
Rubus strigosus Michx.

[494]
A shrub with biennial stems to 2 m tall, sparsely to densely armed with spines. Leaves compound, leaflets 3–5, softly gray-hairy beneath. Flowers in clusters of 2–5, pedicels with stalked glands. Petals white. When ripe the fruits of this species are soft and red, a cluster of juicy drupes that readily fall intact from the whitish receptacle. Abundantly gathered from the wild for food. Parent of the domesticated varieties. Quite variable. Sometimes treated under R. *idaeus* L. as one of several varieties. Occasional. Dry to moist places, roadsides, thin woods, clearings, thickets near stable dunes, bog margins; NY–Nfld. May–Aug.

Rugose Rose
Rosa rugosa Thunb.

[495]
Roses are shrubs, sometimes viny, usually prickly, with alternate pinnately compound leaves. They have

well-developed globose to urn-shaped hypanthiums that conceal the many separate ovaries inside. The stigmas protrude through a small hole in the top. Fruits are usually fleshy and called hips.

The rugose rose is a coarse shrub, usually growing in dense clumps. Twigs with straight prickles and finely hairy. Leaflets dark green, deeply and abundantly furrowed on the upper surface giving it a rough (rugose) appearance. Sepals entire, long persistent. Petals red, pink, or white. The mass of stigmas is barely protruding at the center of the flower. Common. Foredunes, dune sands, peat at margins of salt marshes, rocky shores; MD–NS. June–Sept.

R. eglanteria L. is also a coarse dense shrub. Armed with stout curved prickles. Leaflets 5–9, aromatic, glandular-hairy. Sepal margins usually cut and bearing stalked glands. Rare. Thickets, roadsides, open places, along shores; CT–NS. May–July.

Pasture Rose
Rosa virginiana Mill.

[496]
Plants to 2 m tall. Prickles usually confined to nodes. Leaflets 5–11, glossy above, teeth averaging about 1 mm high. Stipules widening towards tip. Sepals entire, usually conspicuously elongated, finally falling from the red fruits. Styles separate but compacted at the opening in center of flower. Common. Marshes, shores, heaths, interdunes, thickets; NJ–Nfld. May–Aug.

R. carolina L. is similar but the stipules are stiffer and their sides parallel. Leaflets are dull above. Prickles are nearly straight, not broad at base. Common. Dry places, thin woods or in the open, stable dunes, prairies, old fields; nVA–NS. May–July.

Swamp Rose
Rosa palustris Marsh.

[497]
Plants to 2.5 m tall, spreading by rhizomes. Prickles curved, usually confined to nodes, broad at base. Leaflets 5–9, dull above. Styles separate and compacted at the opening in center of flower. Common. Swamps, shores, saw-grass marshes, wet thickets, depressions between stable dunes; NC–NS. June–Aug.

R. laevigata Michx. (Cherokee Rose) is easily recognized. It is a rapid grower and often high-climbing. Leaflets 3, shiny, evergreen, glabrous, often with prickles on larger veins. Stems strongly armed with prickles. Petals white, 3–4 cm long. Hypanthium bristly. Rare generally, but common around buildings, and old residential sites; eFL–SC. Mar.–May.

Black Cherry
Prunus serotina Ehrh.

[498]
Prunus species are shrubs or trees with alternate simple leaves that are more than 2-ranked. Lenticels on trunk and branches are horizontally elongated. True terminal bud present. Fresh foliage and twigs with a bitter-almond taste and odor when crushed and broken. Petals white to pink. Pistil 1, simple, the sides free from the cup-shaped hypanthium that falls soon after petals do except for *P. serotina*. Fruit a drupe. Leaves, twigs, bark, kernals from drupe pits of most, if not all, species can cause cyanide poisoning when eaten uncooked.

Black cherry is a tree to 25 m tall. Leaf margins crenate to crenate-serrate, teeth 6–8 per cm at midmargin. Inflorescence a raceme on tip of new leafy twigs. Pedicels 3–6 mm long. Hypanthium and calyx lobes persisting until after fruit drops. Mature fruit dark purple or black, 7–10 mm across. One of the most poisonous plants in terms of livestock deaths. Common. In woods or in the open, scattered almost everywhere by birds, especially common in fence-rows; MS–NS. Apr.–May.

Choke Cherry
Prunus virginiana L.

[499]
Much like black cherry but leaf margins have sharp divergent teeth, pedicels 5–8 mm long, hypanthium and calyx falling early, mature fruit dark red to nearly black and 8–13 mm across. Common. Borders of woods, dunes, thickets, poor soils, swamp edges; NY–Nfld. Apr.–May.

Pin Cherry; Fire Cherry
Prunus pensylvanica L. f.

[500]
Trees to 12 m tall. Leaves toothed to base, the tips acute to acuminate. Flowers appearing with the leaves, 2–5 in umbel-like clusters from old wood. Sepals glabrous and without marginal glands. Fruit globose, 5–8 mm across, shiny, bright-red, the stone 4–5 mm long. Common. Often abundant after fires. Dry to moist woods, clearings, edge of rocky beaches, stable dunes, thickets, roadsides; NY–Nfld. Apr.–May.

Hog Plum; Black Sloe
Prunus umbellata Ell.

[501]
Large shrub or small tree to 6 m tall. Leaf margins crenate to crenate-serrate, 8–10 glandless teeth per cm at midblade. Flowers in umbel-like clusters, appearing before leaves. Fruit light purple to nearly black, heavily glaucous, 10–13 mm long, their pedicels 10–15 mm long, pulp sour. Common. Dry soils—dunes, thin woods, fencerows, scrub areas; MS–sNC. Mar.–Apr.
   P. maritima Marsh. (Beach Plum) is a straggling shrub to 2.5 m tall, often with some branches on the ground. Leaves with acute tips, underside hairy at maturity, margins with pointed glandless serrations. Flowers as in hog plum. Fruit bluish purple to dark purple, glaucous, 13–25 mm across, highly prized for jellies and jams. Common. Dunes, sandy soils in the open; VA–sME. Apr.–June. *P. angustifolia* Marsh. (Chickasaw Plum)

also has flowers as in hog plum. It may be recognized by the small teeth on leaves, about 20 per cm at mid-margin. Fruit yellow to orange or red, 20–25 mm long, sweet to sour or bitter. Occasional. Usually in thickets, in the open—fencerows, roadsides, thin woods, stable dunes; MS–wFL; GA–sSC. Feb.–Apr.

Laurel Cherry
Prunus caroliniana (Mill.) Ait.

[502]
Evergreen tree to 12 m tall. Horizontally elongated lenticels on younger trunks and large twigs, but absent from the dark-gray tight smooth bark of older trunks. Stipules or stipule scars present. Leaves evergreen, leathery, glabrous on both sides, margins entire or with widely spaced, small, short, sharp, bristlelike teeth. Flowers in short racemes from wood of previous year. Mature fruit a black dull-surfaced drupe with thin nearly juiceless flesh, tending to persist long after maturation, even until the next flowering period, as in the photograph. Common. In variety of habitats, the seeds being abundantly distributed by birds; stable dunes, live oak woods, fencerows, thickets, maritime woods; AL–sNC. Mar.–Apr.

Gopher-apple
Licania michauxii Prance

[503]
A small shrub to 40 cm tall, with horizontal underground stems and alternate shiny evergreen leaves. Flowers similar to those of *Prunus* species and in terminal clusters. Fruits ovoid to obovoid drupes 2–3 cm long with firm flesh and a terete pit. The mature fruits are rarely seen for they are eaten by a number of animals, but especially by land turtles, which are locally called gophers. Common. Stable dune areas, sandy pinelands, scrub oak sandhills, sand pine habitats; MS–FL. May–June. Syn.: *Chrysobalanus oblongifolius* Michx.

FABACEAE
Bean Family

Huisache
Acacia smallii Isely

[504]
Shrub or small tree to 4 m tall, often with more than 1 trunk. Leaves twice-pinnately compound, veins usually invisible. Stipules are nearly straight spines in pairs, rarely 1 per node or none. Flowers fragrant, very small, in yellow globose heads about 25 cm in diameter. Bracts on the peduncles near the globe of flowers. Fruit nearly terete legume 3–8 cm long, about 1 cm across, tapered at both ends, incurved-beaked at apex, the seeds in 2 rows. Formerly the source of fragrant oils of "French" perfumes. Occasional. Sandy soils, roadsides, edge of marshes, banks of ditches, meadows behind dunes; TX–wFL. Mar.–May. Syn.: *Vachellia densiflora* (Alex. ex Small) Cory.
A. farnesiana (L.) Willd. is nearly identical, except veins of leaflets usually visible and apex of fruit merely obtuse to acute. Rare. Similar places; eFL–GA. Mar.–May. Syn.: *Vachellia farnesiana* (L.) Wight & Arn.

Redbud
Cercis canadensis L.

[505]
Small deciduous tree to 12 m tall with simple alternate palmately veined leaves. Blades "heart-shaped," entire, with 5–9 main veins. Petioles swollen at both ends. Twigs geniculate. Fruits are flat oblong legumes 8–11 cm long. Occasional. Rich woods; wFL; GA–sNC. Mar.–Apr.

Indigo-bush
Amorpha fruticosa L.

[506]
Deciduous shrub to 5 m tall. Leaves pinnately compound. Leaflets 11–27 with an odd leaflet at end, entire, dull green. Flowers in 2 to several spike-like racemes. Sepals united, lobes unequal. Petal 1, the standard; purplish, basally surrounding the stamens and pistil. Stamens 10, all united near base. Fruits glabrous, 6–9 mm long, 2–3 mm broad, dotted-glandular, indehiscent. Occasional, rare north of NC. Live oak and maritime woods, mixed hardwoods; eTX–NY. Apr.–June.

Amorpha fruticosa

Black Locust
Robinia pseudoacacia L.

[507]
Shrub or tree to 24 m tall, often suckering abundantly from roots. Leaves pinnately compound. Leaflets entire, 7–11 with an odd leaflet at tip. Flowers with 5 white petals shaped like those of peas. Calyx lobes 2.5 mm long or less. Stamens 10, the filaments united into a tube. Fruit a flat legume, 5–10 cm long, 10–12 mm broad, usually 3- to 8-seeded. Extensively planted for its durable wood, especially for posts, and for ornamental purposes, but can soon become a nuisance because of its vigorous suckering. Occasional. Woods, thickets, roadsides, behind stable dunes; eTX–NS. May–June.

Rattle-bush
Daubentonia punicea (Cav.) DC.

[508]
Shrub to 3 m tall. Leaves once-compound; leaflets entire, 12–40, even-numbered, one of the tip pair sometimes missing, as in the photograph. Corolla orange-red to purplish red, standard 15–20 mm long. Fruit conspicuously 4-winged, 5–8 cm long. Seeds known to be quite poisonous. Occasional. Edge of marshes, roadsides, waste places, between stable dunes, ditches; eTX–NC. June–

Sept. Syn.: *Sesbania punicea* (Cav.)
Benth.

D. drummondii Rydb. is semi-
shrubby, dying to near ground in
colder winters. Corolla yellow, often
with fine red lines, the standard 12–
15 mm long. Fruit similar. Seeds, and
to a lesser extent leaves, are known to
be poisonous. Common locally.
Sandy soils, salt-spray communities,
scrub pine woods; eTX–wFL. Syn.:
Sesbania drummondii (Rydb.) Cory.

RUTACEAE
Rue Family

Prickly-ash
Xanthoxylum clava-herculis L.

[509, 509a]
Aromatic shrub or tree to 12 m tall
with alternate pinnately compound
leaves that usually bear sharp
prickles. Leaflets shiny green above,
5–19, with an odd leaflet at end.
Sharp prickles on the trunk have
broad pyramidal bases. Male and female flowers on separate plants in
large terminal corymblike cymes.
Fruit a fleshy 1- to 2-seeded follicle.
Reasonably tolerant to saline soils.
Common. Dry woods, sandhills,
dunes, thin pinelands, live oak
woods; LA–VA. Mar.–May.

EUPHORBIACEAE
Spurge Family

Silver-leaf Croton; Beach-tea
Croton punctatus Jacq.

[510]
Shrub to 1.2 m tall, branching above
but not densely so. Entire plant except upper surface of leaves densely
covered with stellate hairs, each cluster of hairs with a reddish spot at
middle. Leaves ovate to ovate-
lanceolate or elliptic, entire, 1–6 cm
long. Male flowers, 1, or less com-

monly 2 or 3 together. Fruit a 3-
carpelled, 3-seeded capsule. Nothing
on the dunes resembles this species.
Common. Dunes, sand flats, loose
deep sands; TX–sVA. Nearly all year.

Chinese Tallow-tree
Sapium sebiferum (L.) Roxb.

[511]
Rapidly growing glabrous tree to 10
m tall, with milky sap. Resembling
poplar trees. Petioles with a pair of
glands on upper side near the blade.
Flowers unisexual, male and female
flowers on same plant. Seeds white, 3
per fruit. Under pressure seeds yield
an oil with some properties of tung
oil that is used in paints. Boiling in
water provides the "tallow" used for
candle-making in China. Here grown
as an ornamental, especially for the
foliage, which turns bright red in the
autumn. It can become quite weedy.
Common. Low swampy areas, waste
places, stable dunes; TX–sSC. May–
June. Syn.: *Triadica sebifera* (L.) Small.

EMPETRACEAE
Crowberry Family

Broom Crowberry
Corema conradii (Torr.) Torr. ex Loud.

[512]
Bushy and erect shrub, mostly mat-
like, but to 60 cm tall, forming dense
extensive mats. Leaves linear,
whorled, evergreen, 3–6 mm long.
Flowers in terminal clusters, unisexual, the male and female on separate
plants. The male plants are beautiful
in flower because of the brownish-
purple anthers and purple filaments.
Fruits globose drupes, usually with 3
nutlets, dry when ripe, 1.5 mm wide.
Common. Open sandy ridges, rocky
places, headlands, dunes; NY–Nfld.
Mar.–May.

The similar *Empetrum nigrum* L.
(Crowberry) occurs on eastern Long
Island, NY. Its flowers are solitary in
leaf axils, nearly sessile. Fruit a black
globose berrylike drupe, 4–7 mm

long, and with 6–9 seedlike nutlets. Rare. Similar places; eNY; nME–Greenl. June–July.

Rosemary
Ceratiola ericoides Michx.

[513]
Erect much-branched dense shrub to 2.5 m tall. Young twigs hairy. Leaves linear, 5–15 mm long, whorled, sessile. Flowers sessile in leaf axils. Sepals, petals, and stamens 2. Anthers reddish brown. Fruits are red to olive drupes 2–3 mm across, with 2 red seedlike nutlets 1.5–2 mm long. Highly susceptible to fire and vehicular traffic. Rare. Scrub oak woods, dry open pinelands, stable dunes; MS–FL. Oct.–Nov.

ANACARDIACEAE
Cashew Family

Winged Sumac
Rhus copallina L.

[514]
Shrub or tree to 10 m tall. Stems with dense very short hairs that often wear off with age. Pith large. Leaves pinnately compound. Leaflets 7–17, entire or crenate to rarely serrate. Leaf rachis winged, sometimes quite narrowly. Inflorescence terminal and in fruit drooping at tip. Fruit a densely short hairy drupe 3–4 mm broad, at maturity dull red but persisting and turning brownish. Common. Stable dunes, sandy flats, grasslands, thickets, thin woods, fencerows, edge of brackish and salt marshes; MS–sME. July–Aug.

Staghorn Sumac
Rhus typhina L.

[515]
Shrub or tree to 10 m tall with hairy twigs, petioles, and leaf rachises. Pith of young twigs relatively large. Leaves alternate, once-pinnate. Leaflets 9–31, sessile, serrate, glabrous, glaucous beneath, apex acuminate.

Leaf rachis not winged. Inflorescence terminal, erect. Fruit a bright red densely long-hairy drupe 3–5 mm broad. This species and others of the genus with terminal inflorescences are not poisonous to touch. Common. Fields, thickets, roadsides; heath behind dunes; NY–NS. May–July.
R. glabra L. (Smooth Sumac) is very similar from a distance but the twigs, petioles, and leaf rachises are glabrous. Also the fruits are only densely short-hairy. Occasional. Well-drained soils in the open—stable dunes, fencerows, old fields, woods margins; NJ–sME. June–July.

Poison-ivy
Toxicodendron radicans (L.) Kuntze

[516]
Very poisonous on contact to some people, causing allergic skin reactions. A shrub to 1 m tall, erect to trailing, or most often high-climbing by aerial roots, attaining a main stem width of 15 cm. Leaves alternate, with 3 leaflets, the terminal one conspicuously stalked, the lateral leaflets asymmetrical. Leaflets may be unlobed or lobed, or coarsely serrate. Inflorescences axillary. Fruits are grayish white glabrous drupes 4–6 mm broad. Common. In a great variety of habitats including fencerows, woods, dunes, beaches, brackish swamps and marshes; TX–sNS. Mar.–June. Syn.: *Rhus radicans* L.

Poison Sumac; Thunderwood
Toxicodendron vernix (L.) Kuntze

[517]
Very poisonous on contact to some people. A glabrous shrub or tree to 9 m tall. Bark smooth, usually gray, sometimes to slate-colored. Sap clear, very toxic. Twigs reddish when young, turning gray and bearing many light-orange lenticels. Leaves pinnately compound. Leaflets 7–13, entire, elliptic to oblong or oblanceolate, apex acute to acuminate. Petioles, leaf rachises, leaflet stalks reddish to maroon. Leaflets dull green but in autumn turning to a handsome

orange to scarlet color at which time poisoning often occurs when the branches are picked for display in homes. Inflorescences are panicles arising from leaf axils. Fruits are smooth whitish drupes 4–5 mm in diameter. Occasional. Swamps, marsh borders, seepage areas, other wet habitats; MS–wFL; DE–ME. May–July.

CYRILLACEAE
Cyrilla Family

Spring Titi; Buckwheat-tree
Cliftonia monophylla (Lam.) Britt. ex Sarg.

[518]
Shrub or tree to 6 m tall, nearly always growing in dense stands. Leaves alternate, glabrous, evergreen, leathery, 25–50 mm long; leaf margin entire; netted-vein pattern indistinct on upper surface. Flowers white, borne in terminal racemes. Fruits 3-angled or uncommonly 2 or 4, shiny yellow, turning brown during winter, many persisting until flowers appear the next year, occasionally a few until later. Occasional. Swamps, depressions, seepage areas; MS–wFL. Mar.–Apr.

Summer Titi
Cyrilla racemiflora L.

[519]
Shrub or tree to 9 m tall. Twigs with continuous pith and no stipules or stipule scars. Leaves alternate, glabrous, semi-evergreen, some falling throughout the winter, but a few (sometimes scattered) remaining until new leaves appear; leaf margins entire; netted-vein pattern conspicuous on upper surface. Flowers white, borne mostly in clustered racemes at upper end of previous year's twig. Fruits drupelike, brownish, ovoid, 2–3 mm long. Occasional. Low pinelands, swamps, pond margins; MS–VA. May–July.

AQUIFOLIACEAE
Holly Family

American Holly
Ilex opaca Ait.

[520]
Hollies are shrubs or trees with simple, alternate, pinnately veined, deciduous or evergreen leaves. Fruits are berrylike drupes. Of much help in recognition are the small to minute triangular stipules that turn brown to black and persist for over a year. They may be seen in the photograph of *I. opaca.*

American holly is an evergreen tree to 15 m tall. Leaves coriaceous, evergreen, 2–5 cm wide, spine-tipped, and usually with 1 to many spine-tipped teeth on margins. Flowers unisexual or rarely perfect, axillary. Fruits red, or rarely yellow. Twigs with leaves and fruit are frequently used for decorations at Christmas time. Common. Maritime and live oak woods, deciduous or mixed evergreen-deciduous woods, stable dunes, hammocks; MS–MA. Apr.–June.

Yaupon
Ilex vomitoria Ait.

[521]
Shrub or small tree to 8 m tall. Leaves evergreen, shiny above, the margins finely crenate. Flowers unisexual or rarely perfect, axillary. Calyx lobes, petals, and stamens 4. Mature fruit red or uncommonly yellow. Common. Swamps, wet woods, maritime and live oak woods, stable dunes, sand flats, edges of brackish and salt marshes; TX–VA. Mar.–June.

Inkberry; Bitter Gallberry
Ilex glabra (L.) Gray

[522]
Rhizomatous shrub to 3 m tall, forming extensive colonies. First-year twigs minutely hairy. Leaves with widely spaced crenate "teeth"

Ilex coriacea

towards apex, glabrous, shiny above, lower surface glandular-dotted. Petioles minutely hairy. Mature fruit black, glabrous, 5–7 mm wide. Common. Pinelands, savannas, bog margins, wet thickets, sandy flats; AL; GA–NS. May–July.

I. coriacea (Pursh) Chapm. (Sweet Gallberry) is similar but often taller, to 5 m, and with leaf blades entire or with a few very short spreading bristlelike teeth. Rare. Similar places; MS–wFL; NC. Apr.–May.

Black-alder; Winterberry
Ilex verticillata (L.) Gray

[523]
Shrub or small tree to 5 m tall. Petioles finely hairy. Leaves deciduous, alternate; blades elliptic to obovate, sharply serrate, veins on upper surface prominently depressed. Flowers in axillary peduncled clusters of 1–3. Calyx lobes and petals, 6–8. Calyx lobes obtuse, ciliate. Mature fruits bright red, 5–7 mm wide. Nutlets about 3 mm long, smooth on back. Common. Swamps, wet woods, bogs, depressions between stable dunes, edge of brackish marshes; VA–Nfld. Apr.–July.

I. ambigua (Michx.) Torr. (Carolina Holly) is similar but grows in quite dry habitats; the leaf blades crenate-serrate or uncommonly to sharply serrate, veins on upper surface scarcely depressed; nutlets 4–7 mm long, with furrowed backs. Occasional. Thin dry pinelands, scrub oak, pine-palmetto; AL–sNC. Apr.–June.

Dahoon Holly; Cassena
Ilex cassine L.

[524]
Evergreen tree to 10 m tall. Branches mostly ascending at angles less than 45° with the branch from which they arise. Lower surface of leaf blades not glandular-dotted; blades entire or, if toothed, these absent from at least the lower ⅓ of blade, the teeth widely spaced, very short, sharp, divergent, and appearing as though set onto a smooth margin. Fruit red, rarely yellow, 5–8 mm wide. Occasional. Seepage areas, pond margins, depressions in flatwoods, banks of brackish marshes, ditches; LA–sNC. Apr.–June.

ACERACEAE
Maple Family

Red Maple
Acer rubrum L.

[525]
Deciduous tree to 30 m tall. Leaves opposite, palmately veined and 2- to 5-palmately lobed, rarely unlobed, the lobes cut less than halfway to midvein, the margins serrate. Flowers red to yellow, in axillary clusters, appearing before the leaves. Fruits red to yellow, 2-winged, ripening in the spring. In the fall leaves turn a range of colors from yellow to red. In the southern part of its range leaves of some plants are densely short-hairy beneath and little or not at all lobed. Common. Usually in moist soils, in woods or in the open, edge of marshes, swamps, stable dunes, uplands; eMS–Nfld. Jan.–May.

A. saccharum Marsh. (Sugar Maple) is a tree to 30 m tall, with lobes or large teeth only and no serrations on leaf margins. Flowers appear with the leaves. Fruits ripening in the fall, greenish. Rare. Rich woods; NY–NS. Apr.–May.

[See p. 374 for illustrations of leaves.]

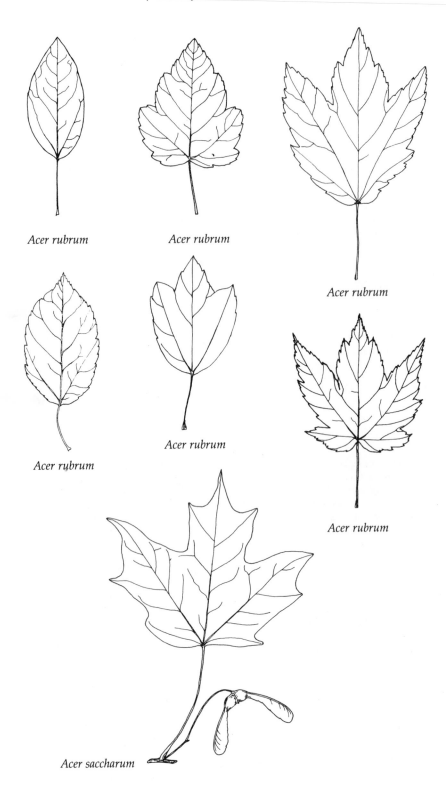

Acer rubrum

Acer rubrum

Acer rubrum

Acer rubrum

Acer rubrum

Acer rubrum

Acer rubrum

Acer saccharum

HIPPOCASTANACEAE
Buckeye Family

Red Buckeye
Aesculus pavia L.

[526]
Shrub or small tree to 10 m tall, often flowering when quite small. Leaves deciduous, opposite, palmately compound. Leaflets 5, rarely more. Conspicuous when in flower. Petals scarlet, stamens longer than lateral petals. Fruit a leathery capsule with 1–3, rarely more, large smooth seeds that may be mistaken for chestnuts. They are poisonous. Occasional. Moist deciduous woods, swamp margins, barrier islands, hammocks, shell mounds; LA–AL; GA–sNC. Mar.–May.

RHAMNACEAE
Buckthorn Family

Rattan-vine; Supple-Jack
Berchemia scandens (Hill) K. Koch

[527]
Twining vine with main stem to 18 cm thick, climbing to tops of tall trees. Leaves alternate; pinnately veined; the lateral veins nearly straight, evenly spaced, and parallel. Flowers greenish, about 2 mm across. Fruit an oblong-ellipsoid drupe, 5–7 mm long, glaucous and blue-black when mature. Common. Usually in moist to wet soils in live oak woods, pine-hardwoods, swamps; LA; FL–VA. Apr.–May.

Buckthorn
Sageretia minutiflora (Michx.) Trel.

[528]
A thin-stemmed elongate weak shrub with many short thornlike branches, often draping over other shrubs and small trees, but not a vine. Leaves opposite or nearly so, 3-veined from near the base. Flowers in spikes 1–2 cm long, calyx about 1 mm long, petals shorter. Fruit purplish black, wingless, drupe-like, 5–9 mm thick, ripening in the spring. When in flower very fragrant and abundantly visited by insects. Occasional. Calcareous hammocks and bluffs, stable sand and shell dunes, shell mounds; MS–sSC. Aug.–Oct.

VITACEAE
Grape Family

Summer Grape; Pigeon Grape
Vitis aestivalis Michx.

[529]
Woody vine, trailing, or climbing by means of tendrils that are usually branched and arise opposite leaves. Bark of older stems shredding. Internodes of stems of current year usually lacking hairs. Year-old stems having a soft brownish pith with hard cross partitions at nodes. Leaves alternate, simple, lobed or unlobed; if lobed, the sinuses rounded at base, coarsely but sometimes shallowly toothed, the underside glaucous or light-colored and nearly glabrous to matted-hairy but not completely concealing the leaf surface. Flowers in panicles 5–15 cm long. Fruit dark-purple to black berries 5–12 mm across. Common. Rich woods, roadsides, thickets, dunes, edge of brackish marshes, fencerows; eTX–MA. May–June. Syn.: *V. rufotomentosa* Small; *V. simpsonii* Munson; *V. bicolor* Le Conte.

Fox Grape; Plum Grape
Vitis labrusca L.

[530]
Leaves and tendrils similar to those of *V. aestivalis* but the lower leaf surface is completely hidden by tawny to reddish matted hairs. Flowers are in panicles 4–8 cm long. Mature berries are dark red to purplish black, 15–25 mm thick. Common. Rich woods, wet to dry thickets, margin of woods, bogs, dunes, roadsides; VA–sME. May–June.

V. mustangensis Buckl. (Mustang Grape) is similar but hairs are all light gray and the mat of cobwebby hairs may be separated with a needle from the very short hairs beneath. Rare. Similar places; TX–LA. May–June.

Muscadine; Scuppernong
Vitis rotundifolia Michx.

[531]
Also called Bullace Grape. Woody vine, trailing or climbing by tendrils that are never branched and arise opposite the leaves. Bark not shredding. Pith lacking a hard diaphragm at the nodes. Fully expanded leaves dentate, unlobed, or shallowly lobed. Flowers in panicles 2–4 cm long. Berries few, in a tight cluster, black, not glaucous, 10–25 mm thick, ripening from Aug.–Oct. Two types of domestic grapes originated from this species. Those with amber-green fruits are known as scuppernongs while those with purple fruits are muscadines. Plants withstand deep sand burial but are susceptible to salt-water spray. Common. Hammocks, dunes, sand flats, brackish marshes, swamps, live oak and maritime woods; eTX–DL. May–June.

Some manuals segregate plants of this species having berries 10–15 mm in diameter and with tender skin and pulp as *V. munsoniana* Simpson ex Munson, but the 2 forms appear to intergrade to the extent that they should not be recognized formally.

Virginia-creeper; Woodbine
Parthenocissus quinquefolia (L.) Planch.

[532]
Woody vine climbing by tendrils tipped with adhesive discs, sometimes shrubby or trailing. Leaves alternate, palmately compound. Leaflets 5, rarely some 3, 6 or 7; finely to coarsely serrate above an entire basal part. Sometimes called "Five-leafed Ivy" but not closely related. Leaves turn rich reddish colors in autumn. Fruits are purple to black drupes, glaucous, globose, 5–9 mm thick. Common. Rich or poor woods, around buildings, stable dunes, maritime forests, edges of freshwater or brackish marshes, hammocks; eTX–sME. May–Aug.

Pepper-vine
Ampelopsis arborea (L.) Koehne

[533]
Woody vine with alternate twice-compound or sometimes partly 3-times compound leaves. Often high-climbing by means of scattered tendrils. Leaflets with conspicuous triangular teeth. Flowers in peduncled clusters that are usually attached opposite the leaves. Petals 5, greenish-yellow, 1.5–3 mm long. Fruits are black berries 6–10 mm thick. Common. Moist woods, marshes, alluvial soils, swales, sandy flats, pond margins, wet thickets; TX–sVA. June–Aug.

HYPERICACEAE
St. John's-wort Family

St. Andrew's-cross
Hypericum hypericoides (L.) Crantz

[534]
Hypericum species are shrubs or herbaceous plants with entire opposite leaves. The blades ordinarily have transluscent internal glands visible with transmitted light, but a hand lens may be needed to see them. Sepals and petals 4–5 and separate, the latter usually yellow. Stamens are numerous in seaside shrubs. Ovary superior, hypanthium absent. Fruit a capsule, 1-celled, 2- to 5-carpelled, with many seeds. Some species cause sensitivity to light when eaten.

This species is an erect, much-branched shrub to 1.5 m tall. Leaves sessile. Flowers and fruits on short erect pedicels. Petals 4. Sepals 4, of which 2 are much smaller than the others. Styles and carpels 2. Common. Dry to moist places, woods between established dunes, sandy flats, edge of marshes, swales, pond margins; eTX–VA. May–Sept. Syn.: *Ascy-*

rum hypericoides L.

H. suffruticosum Adams & Robson has similar characteristics except it is usually decumbent, only to 15 cm tall, and has long pedicels that become reflexed soon after petals fade. Rare. Dry sandy thin soil, pinelands; eLA–sGA. Apr.–June. Syn.: *Ascyrum pumilum* Michx.

St. John's-wort
Hypericum tetrapetalum Lam.

[535]
A shrub to 90 cm tall with 4 sepals, 2 quite small, and 4 petals. Leaves ovate to oblong-ovate, cordate, clasping the stem, glandular-dotted on both surfaces. Carpels 3. Styles diverging. Occasional. Sandy or peaty soils in pine flatwoods, pond margins, marshes, depressions, roadside ditches; FL–sGA. Mar.–Sept.

H. crux-andreae (L.) Crantz (St. Peter's-wort) has similar flowers, but the leaves are sessile and not clasping, the blades elliptic to ovate or obovate. Occasional. Dry to moist pinelands, bogs, meadows, shores of ponds; eTX–NY. June–Oct. Syn.: *H. stans* (Michx.) Adams & Robson, *Ascyrum stans* Michx. ex Willd.

Hypericum crux-andreae

Hypericum tetrapetalum

H. myrtifolium Lam. is much like St. John's-wort vegetatively, but only the lower leaf surface is glandular-dotted. Sepals 5, nearly equal. Petals 5. Styles erect. Common. Pond margins, cypress ponds, pine flatwoods; eMS–GA. May–Aug.

St. Peter's-wort
Hypericum cistifolium Lam.

[536]
Erect shrub to 90 cm tall. Stems simple or with relatively few branches. Leaves lanceolate to linear-lanceolate or linear-oblong, 4–10 mm wide, 1.5–4 cm long. Flowers many in seemingly naked terminal clusters. Petals and sepals 5. Seeds 0.5–0.7 mm long, fastened on inside of ovary wall. Common. Sandy soils—swales, pine flatwoods, seepage areas, bogs, roadsides, margins of swamps and marshes; eLA–sNC. Apr.–Oct.

H. fasciculatum Lam. (Sand-weed) also is an erect shrub, to 2 m tall, much branched above. Bark roughened and slightly spongy, peeling in thin sheets. Young stems strongly flattened, winged. Leaves linear-subulate or needlelike, the largest 13–26 mm long. Leaves of the short axillary branches often appearing fascicled in the axils. Occasional. Edges of cypress ponds and other ponds, wet pinelands, ditches, swales; MS–sNC. May–Sept.

Hypericum cistifolium

TAMARICACEAE
Tamarisk Family

Tamarisk; Salt-cedar
Tamarix spp.

[537]
Shrubs or small trees with the small leafy twigs resembling forms of *Juniperus virginiana*, the Red-cedar. Leaves alternate, evergreen, scalelike, not divergent. Flowers small, short-pedicelled or sessile, mostly on green branches, in spikelike racemes that are usually clustered so as to form a panicle. Sepals 5. Petals 5, white to pink and persisting. Stamens 5 or rarely more, arising from the edge of a disc. Ovary 1-celled. Fruit a capsule with many seeds. Tip of seeds with a tuft of hairs. Occasional. Introduced and escaping, sandy soils, dunes, meadows, old fields, saline or brackish areas; TX–sNC. Mar.–Oct.

There are probably 3 species within the range of this book, but identification is hard. *T. chiensis* Lour., the most common, has sepals entire, or nearly so; petals persistent. In *T. gallica* L. and *T. canariensis* Willd., petals drop soon after flowering; in the latter, sepals have minute teeth, not so the former.

CISTACEAE
Rock-rose Family

Rock-rose; Sun-rose
Helianthemum corymbosum Michx.

[538]
Shrub to 35 cm tall, often growing in dense colonies. Roots tuberous-thickened. Stem leaves 10–20, elliptic to oblanceolate, 1.5–3 cm long, 5–10 cm wide, loosely stellate-hairy above, densely hairy beneath. Petal-bearing flowers in dense terminal clusters, on pedicels 1–2 cm long, petals 5 and about 1 cm long. Stamens many. Stigmas globose, prominent. Petal-bearing flowers mixed with nonopening flowers, these on short pedicels,

with only 3–8 stamens, and in compact terminal clusters. Fruit a glabrous 3-celled capsule 5–7 mm long, with few seeds. Occasional. Dunes, thin maritime woods, dry scrub; MS; FL–NC. Feb.–May.

H. arenicola Chapm. is shrubby (herbaceous in very cold winters), low and spreading, with flowers scattered in loose clusters. Longest sepals 6–8 mm. Ovary and capsule densely stellate-hairy. Rare. Dunes, edge of salt marshes; MS–wFL. Mar.–Apr. *H. georgianum* Chapm. is similar but the sepals are 4–6 mm long. Fruits 4 mm long. Rare. Maritime forest, dunes, old fields; eTX–NC. Apr.–June.

Hudsonia; Beach-heather
Hudsonia tomentosa Nutt.

[539]
Spreading, low, much-branched, groundcover shrub, often forming extensive colonies. Leaves evergreen, alternate, crowded, closely appressed, ovate-lanceolate, 1–4 mm long. Flowers on pedicels 1–5 mm long, usually numerous. Petals 5, yellow. Stamens 5–25. Style slender, elongated. Fruits are glabrous ovoid 1-celled capsules with 1–2 seeds, surrounded by the persistent sepals. Common. Dunes, loose sands, sandy flats, sandy meadows; nNC–NS. May–July.

H. ericoides L. (Golden-heather) has a very similar aspect but the leaves are linear, 2–7 mm long, erect or ascending. Flowers on slender naked pedicels 4–10 mm long. Capsule cylindrical, hairy at tip. Occasional. Dry sands, thin pinelands, dunes, upper beaches, dry open flats; DE–Nfld. May–July.

CACTACEAE
Cactus Family

Eastern Prickly-pear
Opuntia humifusa (Raf.) Raf.

[540]
Opuntia species stems are perennial,

of segments that are large, thick, and succulent. Segments, also called joints, may lack large spines but more often they are present and obviously to be avoided. In addition there are circular areas (areoles) with tufts of numerous tiny barbed bristles that can penetrate skin easily. Therefore, extreme caution should be used in handling the plant even in absence of large spines. Leaves are small, fleshy, pointed, fastened below the areoles, and fall early. Fruits are berries, the seeds flattened.

O. humifusa grows in clumps or is mat-forming. Segments hold together tightly, are 6–9 mm thick, 4–16 cm long, 4–12 cm broad, and with or without spines. Areoles are about 3 mm across and 1–2 cm apart. Common. Dry open areas or thin woods—dunes, sandy flats, pastures, fence-rows; TX–sMA. Apr.–June. Syn.: *O. compressa* (Salisb.) Macbr.

O. stricta (Haw.) Haw. (Southern Prickly-pear) is similar but joints are 12–20 mm thick, 10–40 cm long, 7.5–25 cm broad. Areoles are 4.5–6 mm across and 3–5 cm apart. Occasional. Similar places; eTX; eLA–MS; eFL; sSC. Apr.–June. *O. ficus-indica* (L.) Mill. (Indian-fig) is erect and develops a trunk. Fruits are 5–10 cm long and 4–9 cm thick. Rare. Similar places; FL–GA.

Devil-joint
Opuntia pusilla (Haw.) Haw.

[541]
Creeping, often mat-forming plants. Stem segments easily detached, nearly cylindrical to slightly flattened, 1–5 cm long, and bearing many needle-sharp, strongly barbed spines to 3 cm long. Stem segments are often seen adhering to one's clothing or skin before they are seen on the ground. Common. Dunes, thin sandy woods, fencerows, pastures, road-sides; eTX; AL–NC. May–June. Syn.: *O. drummondii* Grah., *O. tracyi* Britt.

LYTHRACEAE
Loosestrife Family

Water-willow; Swamp Loosestrife
Decodon verticillatus (L.) Ell.

[542]
Weak-stemmed, branches strongly arching, often rooting at ends. Lower part of main stem exfoliating in long cinnamon-colored strips. Bark under water or near water line thick and spongy. Terminal parts of plant may die during cold winters. Young stem densely hairy. Leaves opposite or whorled, with short petioles, blades lanceolate to elliptic-lanceolate, acute at both ends. Flowers in axillary clusters. Hypanthium globose to bell-shaped, 4.5–5 mm long. Fruits are dark brown capsules 4–7 mm thick and closely surrounded by the persistent hypanthium, often remaining into next growing season. Occasional. Pond margins, sloughs, freshwater and brackish marshes, open areas in swamps, ditches; eFL–NS. July–Sept.

ARALIACEAE
Ginseng Family

Hercules'-club; Devil's-walking-stick
Aralia spinosa L.

[543]
Shrub or tree, usually little-branched, to 10 m tall, with coarsely prickly stems and leaves. Leaves several times compound, quite large, leaflets many. Flowers in numerous umbels, arranged in a large terminal panicle. Conspicuous when in flower, flowers about 5 mm wide. Fruits berrylike drupes 4–6 mm across, black, with 5 stony pits. Insects in large numbers visit flowers. The fruits are frequently stripped from the plants by birds. Occasional, prominent when in flower or fruit. Thin or dense woods or in open, moist or dry places—hammocks, stable dunes, swamp margins; MS–MD. June–Sept.

CORNACEAE
Dogwood Family

Swamp Blackgum
Nyssa biflora Walt.

[544]
Deciduous tree to 40 m tall, some-times plants of low stature and bushy tops in dense colonies. True terminal bud present. Pith continuous but with firm diaphragms. Bundle scars 3. Leaves simple, alternate, not 2-ranked, entire. Leaf blades firm, shiny above, glabrous or nearly so below, oblanceolate to narrowly oblong-lanceolate or obovate, apex obtuse to acute. Fruits are bluish-black ellipsoid to nearly globose drupes about 1 cm long, in clusters of 2, rarely 3, or rarely single, at end of a peduncle to 4 cm long. Vegetatively sometimes confused with persimmon, which has a single bundle scar and lacks a true terminal bud. Common. Swamps, pond margins, seepage areas, edge of brackish and freshwater marshes; eTX–DL. Mar.–June.

N. sylvatica Marsh. (Sourgum) is similar but leaves are usually widest at or near the middle; the apex usu-ally acuminate; fruits in clusters of 3 or 4, sometimes 2, or rarely single on end of a peduncle to 7 cm long. Often treated with *N. biflora*, the two as va-rieties of the same species. Usually well-drained places—woods, fence-rows, old fields; occasionally in moist places; rare eTX–NC; common NY–sME. Apr.–June.

Flowering Dogwood
Cornus florida L.

[545]
Dogwood leaves are simple, decidu-ous, and pinnately veined. Venation is distinctive in that the main lateral veins of most leaf blades arise from only the lower ⅔ of the midrib (use a ruler), and at least the uppermost vein on each side is arched inward so that its tip approaches the midvein. Fruits are drupes, developed from an inferior ovary.

C. florida is a deciduous shrub or tree to 9 m tall with a broad much-branched crown. Bark broken into small square plates. Leaves opposite. Flowers in a dense head of 15–30, surrounded by 4 showy white, rarely pink, bracts that develop from the flower bud scales. These scales may be seen in their overwintering condi-tion on the 2 flower buds in the pho-tograph. Petals 4, greenish to yellow-ish, 3–4.5 mm long. Fruits are scarlet, shiny, 8–18 mm thick. An attractive plant, especially when in flower in spring and when leaves are brilliant red in fall. Common. Understory tree in woods of almost every type on well-drained soils. Also scattered in open areas, probably growing from fruit pits dropped by birds; AL–sME. Mar.–May.

C. canadensis L. (Dwarf Cornel) has similar heads of flowers but only 1 long-peduncled head per plant, plants only to 30 cm tall, and with a whorl of 4–6 leaves. Mature fruits red. Rare. Woods, bogs; nNJ–Nfld. May–June.

Red-willow; Silky Cornel
Cornus amomum Mill.

[546]
Much-branched deciduous shrub to 3 m tall. Pith of second-year twigs

Nyssa sylvatica   *Nyssa sylvatica*

brown. Leaves smooth above, finely hairy beneath. Mature fruit blue to purple. Occasional. Wet habitats—freshwater marshes, swamps, thickets; VA–NY. June–July.

C. stricta Lam. (Stiff Dogwood) also has blue fruits but pith of second-year twigs is white. Leaves are smooth above, glabrous beneath or with a few fine, short, straight, appressed hairs. Occasional. Similar places; MS–NC. Apr.–May. Syn.: *C. foemina* Mill. ssp. *foemina*.

In *C. asperifolia* Michx. the pith is also white and the fruits are blue but leaves are finely roughened above and the undersides have forked, curling, appressed to spreading hairs. Rare. Shell mounds, wooded stable dunes, low woods, live oak woods; eFL–sNC. May–June. Syn.: *C. foemina* Mill. ssp. *microcarpa* (Nash) J. S. Wils.

Alternate-leaved Dogwood; Pagoda Dogwood
Cornus alternifolia L. f.

[547]
Shrub or small tree to 8 m tall with a flat-topped spreading crown. Leaves deciduous, alternate but many are crowded at ends of twigs and appear whorled, being obviously alternate only on long twigs. Petioles slender, 2–6 cm long. Flowers in open cymes. Mature fruit globose, bluish-black, 4–7 mm across. Occasional. Woods, thickets; nNJ–Nfld. May–June.

CLETHRACEAE
White Alder Family

Sweet-pepperbush
Clethra alnifolia L.

[548]
Deciduous shrub to 3 m tall. Twigs with small stellate hairs. Leaves 4–11 cm long, 25–50 mm wide, oblanceolate to elliptic-oblanceolate, apex acute to short-acuminate, margins sharply serrate but entire toward the base. Flowers in racemes, fragrant.

Fruit a 3-carpelled capsule about 3 mm long, nearly globose. Common; rare along Gulf Coast. Swamps, moist thickets, wet pinelands; eTX–sME. May–July. Syn.: *C. tomentosa* Lam.

ERICACEAE
Heath Family

Lambkill; Sheep-laurel
Kalmia angustifolia L.

[549]
Shrub to 1.7 m tall with glabrous twigs, opposite or whorled evergreen leaves 25–70 mm long. Blades leathery, elliptic to elliptic-lanceolate or oblong, blunt-tipped. Calyx with glandless hairs. Corolla tube much expanded on outer portion, somewhat bell-shaped. In newly opened flowers the 10 anthers are seated in small pockets of the expanded portion of the corolla tube. Ovary superior. Fruit a subglobose 5-carpelled capsule. Occasional. Thin pinelands, wet thickets, swales, dry to moist mixed woods; nVA–Nfld. May–July.

K. hirsuta Walt. (Hairy Wicky), a weak shrub to 60 cm tall, has very similar flowers but leaves are alternate, 5–15 mm long, persistent, blades ovate to oblong or oblanceolate. Twigs with numerous long spreading hairs. Calyx 5–8 mm long, lobe margins conspicuously ciliate. Occasional. Dry to wet places—thin pinelands, bog margins; AL–sSC. May–July.

Tree Lyonia; Stagger-bush
Lyonia ferruginea (Walt.) Nutt.

[550]
Evergreen shrub or tree to 6 cm tall, often in large clumps. Lower leaf surface, pedicels, and calyx bearing rust-colored shieldlike scales that become grayish with age. Flowers and fruits borne on twigs of the previous season at about the time new leafy shoots are developing. Fruit an ovoid, scaly, finely hairy, 5-angled capsule 3–6 mm

long. Common. Pinelands, pine-palmetto woods, live oak woods, scrub pine, oak-pine woods, wet pinelands; FL–sSC. Mar.–Apr.

L. fruticosa (Michx.) G. S. Torr. ex B. L. Robins. is much like the above species but most, if not all, flowers develop from current year's growth. Rare. Wet pinelands; FL–GA. May–June.

He-huckleberry; Maleberry
Lyonia ligustrina (L.) DC.

[551]
Deciduous shrub to 4 m tall. Leaves alternate, margins minutely serrate, blades lanceolate to elliptic or obovate, to 8 cm long, frequently turning red in fall. Flowers and fruits in small umbels usually arranged on racemes, these from the axils of leaf scars on the previous year's twigs. Calyx lobes 0.5–1.5 mm long. Corolla urn-shaped, 2–4.5 mm long, the lobes curved outward. Fruit a nearly globose capsule 2.5–3 mm long. Occasional. Between stable dunes, thickets, depressions, pine flatwoods, pond margins; SC–sME. Apr.-June.

Seaside plants of this species are almost entirely var. *ligustrina*, which has no or only a few small leafy bracts mixed with the flowers. *L. ligustrina* var. *foliosiflora* (Michx.) Fern. with conspicuous leafy bracts mixed with the flowers may be encountered in SC–sVA. Apr.–June.

Fetterbush
Lyonia lucida (Lam.) K. Koch

[552]
Evergreen shrub to 4 m tall, first-year twigs strongly angled. Leaves alternate, 2–8 cm long, rigidly coriaceous, margins entire, with a distinct marginal vein, mature leaves glabrous. Flowers in drooping axillary clusters from axils of leaves of previous year. Calyx lobes persistent, 4–5 mm long, apex acute. Corolla white to pink, 6–10 mm long, the base swollen. Fruit an ovoid capsule 4–5 mm long, with a truncate apex. Common. Wet pinelands, depressions in broadleaf woods, seepage areas; AL–sVA. Apr.–June.

Leatherleaf
Cassandra calyculata (L.) D. Don.

[553]
Evergreen shrub to 1.5 m tall with arching flowering stems. Leaves 15–50 cm long, lower surface with many grayish to brownish peltate scales. Flowers solitary in axils of previous year's leaves. Corolla ovoid, 6–7 mm long, with a small opening at top, lobes 5 and outward-curving. Fruit a capsule about 3 mm wide and 2.5 mm tall. Occasional. Bogs, pond margins; NJ–Nfld. Mar.–May. Syn.: *Chamaedaphne calyculata* (L.) Moench.

Trailing-arbutus
Epigaea repens L.

[554]
Prostrate, creeping evergreen shrub with alternate leathery leaves 3–10 cm long, blades entire and rounded or cordate at base. Flowers fragrant, in small terminal or axillary clusters. Corolla white to pink, densely hairy on inside, lobes 5. Fruit a globose hairy 5-carpelled capsule 5–8 mm long. Seeds many. Rare. Well-drained soils—woods, thickets, open places; VA–Nfld. Mar.–May.

Wintergreen; Teaberry
Gaultheria procumbens L.

[555]
Aromatic shrub with odor of wintergreen, with extensive creeping stems and erect leafy and flowering stems to 15 cm tall. Leaves alternate, evergreen, 2–6 cm long. Flowers 1 to a few, solitary in leaf axils. Corolla slightly bell-shaped to urn-shaped, 6–12 mm long. Fruit 7–10 mm long, berrylike, the enlarged hypanthium forming the fleshy part, often persisting through winter. Occasional. Dry to moist woods, thin scrub; nVA–Nfld. July–Aug.

Bearberry
Arctostaphylos uva-ursi (L.) Spreng.

[556]
Much-branched prostrate evergreen shrub forming dense mats to 1 m across. Leaves alternate; blades entire, oblanceolate to oblong-obovate, 1–3 cm long, apex obtuse to rounded. Corolla white to pink, 4–6 mm long. Fruit a bright red drupe 6–10 mm long, with a pit of 5 adhering nutlets, flesh dry and mealy. Occasional. Well-drained sands, gravel banks, thin scrub, stable dunes; NJ–Greenl. May–July.

Black Huckleberry
Gaylussacia baccata (Wang.) K. Koch

[557]
Erect shrub to 1 m tall with yellowish glandular dots on pedicels and both sides of the leaves. Hairs are absent on these structures and the twigs. Leaves deciduous, entire, 2–5 cm long. Corolla ovoid-conical to nearly cylindrical, 4–6 mm long, the lobes curved outward. Ovary inferior. Fruit black, 6–8 mm long, a berrylike drupe with 10 seedlike pits. Common. Dry places in woods or in the open, thickets, stable dunes; VA–Nfld. May–July.

G. frondosa (L.) T. & G. ex Torr. var. *frondosa* (Dangleberry) is a shrub to 2 m tall also with yellowish glandular dots on pedicels but only on underside of leaves. Hairs are also absent. Fruits are light purple, glaucous, 7–10 mm long. Occasional. Dry to wet places—in woods or in the open, thickets, pinelands; VA–MA. Apr.–June. *G. frondosa* var. *tomentosa* Gray (Hairy Dangleberry) has hairs on twigs and lower surface of leaves. Rare. Usually moist to wet places—thin pinelands, live oak-pine woods; eFL–GA. Mar.–May.

Dwarf Huckleberry
Gaylussacia dumosa (Andr.) T. & G.

[558]
Colonial shrub to 40 cm tall from subterranean runners. Leaves, pedicels, and sepals with small stalked glands.

Bracts in the flowering racemes persistent until fruits mature. Fruit black, usually hairy, 5–8 mm long. Occasional. Bogs, interdunal flats, swales; nNJ–Nfld; rare AL–NC. June–July.

Low Sweet Blueberry
Vaccinium angustifolium Ait.

[559]
Seaside *Vaccinium* species have alternate leaves. The ovary is inferior. Fruits are berries. The species are much like *Gaylussacia* but leaves lack glands. Although blueberries in general may be difficult to identify, the 9 species we have recorded for seaside habitats are relatively easy to recognize.

V. angustifolium is a small, much-branched shrub to 30 cm tall from underground runners and usually colonial. Leaves are bright green on both sides, 4–10 mm wide, 1–3 cm long, finely and sharply serrate. Corolla 3–5 mm long, tubular, nearly cylindrical to ellipsoid. Fruit 5–7 mm thick. Common. Dry places in thin woods or in the open; nNJ–Nfld. May–June.

V. pallidum Ait. (Lowbush Blueberry) is also of low stature, rarely reaching 50 cm. Leaves 15–30 mm wide, 2–6 cm long, entire or finely serrate, pale green beneath. Rare. Dry places—thin woods or in the open; nNC–NS. Mar.–May. Syn.: *V. vacillans* Kalm ex Torr.

Highbush Blueberry
Vaccinium corymbosum L.

[560]
Highly variable shrub 1–4 m tall. Plants glabrous or hairy. Leaves deciduous, entire or serrate, to 8 cm long. Corolla 5–10 mm long; nearly cylindrical to urn-shaped to somewhat bell-shaped; white, yellowish, greenish, pink-tinged, or pink; with short outwardly curved lobes. Fruits dark blue to black, glaucous or not, 6–12 mm long. No other seaside blueberry grows as tall as highbush. Common. Wet to dry places—between stable dunes, pine woods, live oak woods, bogs, swamps, seepage areas;

eTX–NS. Feb.–June. Syn.: *V. atrococcum* (Gray) Heller, *V. elliottii* Chapm., *V. fuscatum* Ait., *V. simulatum* Small.

Sparkleberry; Tree Blueberry
Vaccinium arboreum Marsh.

[561]
Shrub or tree to 10 m tall with thin, light, reddish brown bark covered with fine scales. Leaves coriaceous, veiny, 2–6 cm long, shiny above, entire or with tiny scattered teeth or points, a few persisting through the winter to evergreen in warm areas. Flowers in leafy racemes. Corolla white, bell-shaped, 5–8 mm long, with short lobes. Anthers with a pair of awns as well as the 2 small terminal tubes. Fruit black, glabrous, pulp surrounding the 8–10 seeds scanty but pleasant-tasting, persisting into winter. Common. Dry places—live and maritime oak woods, pinelands, stable dunes, scrub areas; MS–sNC. Apr.–June.

Evergreen Blueberry
Vaccinium myrsinites Lam.

[562]
Erect evergreen shrub to 60 cm tall. Leaves glabrous or nearly so, 8–30 mm long, shiny, with a tiny point on tip, the lower surface with stalked or club-shaped glandular hairs. Corolla white to pink, urn-shaped, 6–8 mm long. Fruit a berry, purple to black, sometimes glaucous, 6–8 mm across. Common. Dry to wet places—pinelands, thin scrub, bog margins, between stable dunes; AL–GA. Mar.–Apr.
  V. darrowii Camp has the same general appearance but the lower leaf surface lacks hairs. Fruit 4–6 mm across. Rare. Similar places; MS–wFL. Mar.–Apr.

American Cranberry
Vaccinium macrocarpon Ait.

[563]
Evergreen trailing shrub with very slender stems. Leaves coriaceous, elliptic-oblong, blunt or rounded at

apex, entire. Pedicels bearing above the middle a pair of green leaflike bracts 2–4 mm long. Corolla with 4 lobes 6–10 mm long and becoming recurved; the tube much shorter. Fruit a bright red globose berry 1–2 cm long, persisting through winter. Occasional. Bogs, fresh marshes, pond shores, swales; MD–Nfld. May–July.
  V. oxycoccos L. (Small Cranberry) has the same general aspect but leaves are ovate-oblong to ovate or triangular, the pedicels with bracts below the middle, and berries 5–8 mm across. Rare. CT–Nfld. June–July.

Deerberry
Vaccinium stamineum L.

[564]
Deciduous shrub to 5 m tall. Leaves 3–10 cm, glabrous or hairy, glaucous or not. Flowers in racemes, from axils of leaves as big as other leaves to as small as 5 mm long. Corolla broadly bell-shaped with 5 short lobes. Anthers with 2 awns as well as 2 tubes. Fruits greenish to yellowish or purplish, solitary in axils of leaves or leaf-like bracts. Occasional. Dry places—pinelands, thin live oak woods, pine-palmetto, mixed hardwoods; AL–VA. Apr.–June.

SAPOTACEAE
Sapodilla Family

Southern Buckthorn
Bumelia tenax (L.) Willd.

[565]
Shrub or tree to 6 m tall. Twigs with silky hairs when young, bearing spines although sometimes scarce. Leaves deciduous in cold winters to evergreen in warmer ones, entire, not 2-ranked, often in fascicles, dull above, covered with nearly parallel appressed silky hairs beneath. Hairs sometimes wear off with age. Fruit a black 1-seeded berry 10–12 mm long, resembling a cherry. Common. Dry places—dunes, pinelands, maritime and live oak woods, mixed hard-

woods; eFL–SC; rare wFL. May–June.

B. *lanuginosa* (Michx.) Pers. (Gum Bumelia) has many similar characteristics but the hairs on the lower surface of the leaf are usually not strongly appressed and tend to spread in several directions. Occasional. Pinelands, mixed hardwoods, between stable dunes; TX–wLA; AL–wFL. June–July.

EBENACEAE
Ebony Family

Persimmon
Diospyros virginiana L.

[566]
Deciduous tree to 21 m tall with bark furrowed into small, square, scaly plates. Twigs with diaphragmed pith, no stipules or their scars, 1 bundle scar, and true terminal bud absent (see photograph). Leaves are whitish beneath and usually have black blemishes on their upper surfaces. Corolla pale yellow, bell- or urn-shaped, to 15 mm long. Fruit a smooth, globose, pale orange berry to 4 cm across with very large flat hard seeds, ripening in late autumn. Wet to dry places—in a variety of woods, ponds, swales, stable dunes, old fields, fencerows, around premises; eLA–NY. May–June.

SYMPLOCACEAE
Sweetleaf Family

Sweetleaf; Horsesugar
Symplocos tinctoria (L.) L'Her.

[567]
Shrub or tree to 6 m tall. Twigs brownish to reddish, usually with an ash-colored surface due to grayish hairs. Pith chambered. Leaves evergreen, or in the north or during severe winters mostly deciduous, entire to serrate-crenate towards apex, not 2-ranked, undersurface with short

hairs, sweet to taste, especially near middle of blade. Flowers from axils of leaves of previous year. Corolla white to yellow. Stamens many, anthers yellow. Plants usually scattered but easily detected at a distance when in flower. Fruits are ellipsoid drupes 8–12 mm long. Occasional. Live and maritime oak woods, mixed hardwoods, stable wooded dunes; MS–sDE. Mar.–May.

OLEACEAE
Olive Family

Devilwood; Wild-olive
Osmanthus americanus (L.) Benth. & Hook. f. ex Gray

[568]
Evergreen shrub or tree to 9 m tall. Bark gray, thin, scaly. Dropped scales expose the reddish inner bark. Leaves opposite, margins entire and rolled inward, glabrous beneath, tip acute to rounded, base acute, 8–20 cm long, usually 2.5 times as long as wide. Flowers small, greenish to light cream, individually inconspicuous, in short axillary clusters, the immature clusters evident late in the growing season and during winter. Mature fruits dark purple ovoid to ellipsoid drupes 10–15 mm long. Common. Dry places—stable dunes, live oak and maritime woods, pinelands; eLA–sVA. Mar.–Apr.

Florida Privet
Forestiera segregata (Jacq.) Krug & Urban

[569]
Evergreen shrub or tree to 6 m tall with smooth light-gray bark. Leaves opposite, 2–5 cm long, glabrous, margins entire, tip rounded, and very finely glandular-dotted beneath. Flowers inconspicuous, greenish yellow, in small clusters from axils of previous year's leaves. Fruit a dark purple to black ellipsoid drupe, often glaucous, 5–11 mm long. Occasional. Shell mounds in marshes and along

shores, dunes, scrub; east part of wFL–sSC. Sept.–Apr. Syn.: *F. porulosa* (Michx.) Poir.

LOGANIACEAE
Logania Family

Yellow Jasmine
Gelsemium sempervirens (L.) St.-Hil.

[570]
Twining woody vine with opposite, evergreen, leathery leaves. Blades glabrous, entire. Flowers fragrant, in small clusters from axils of leaves of previous year. Corolla yellow, tubular, 2–4 cm long. Fruit a 2-celled capsule, the main body 14–23 mm long, 8–11 mm wide, flattened contrary to the partition. Common. Stable dunes, thickets, fencerows, thin woods; FL–sVA. Jan.–Mar.

　G. rankinii Small (Swamp Yellow Jasmine) has almost the same characteristics. It may be distinguished by having odorless flowers and the body of the fruit only 9–12 mm long. Rare. Wet places in woods or open; AL. Feb.–Apr.

VERBENACEAE
Vervain Family

Shrub-verbena; Lantana
Lantana camara L.

[571]
Weak shrub to 1.6 m tall. Stems armed with scattered prickles, young stems squarish. Leaves deciduous, opposite, scabrous, especially above. Blades 2–7 cm long. Flowers in heads, surrounded by bracts much shorter than the flowers. Corolla tubular; with 4 unequal lobes; at first cream, yellow, or pink; usually changing to orange or scarlet with age. Ovary superior. Mature fruits are dark blue to black drupes about 8 mm long, with 2 bony nutlets. Occasional. Mostly planted but abundantly naturalized in

waste places and around buildings, and scattered in the wild; TX–GA. May–Sept.

Beauty-berry; French-mulberry
Callicarpa americana L.

[572]
Deciduous shrub to 5 m tall with arched stems. Stems slender, round to slightly 4-sided, finely stellate hairy. Leaves opposite, sometimes a few only nearly opposite, 8–20 cm long, margins coarsely serrate, tips acute to acuminate, slightly to densely hairy. Flowers in short axillary clusters, small. Petals white to pink, united, with lobes 3–5 mm long. Fruit a juicy berrylike drupe with 4 small nutlets, violet to magenta, or white in a rare form. Common. Live oak and maritime woods, mixed hardwoods, stable dunes, fencerows, shrub thickets; TX–VA. June–Aug.

AVICENNIACEAE
Black Mangrove Family

Black Mangrove
Avicennia germinans (L.) L.

[573]
Evergreen shrub to 5 m tall. Young twigs with dense fine hairs. Leaves opposite, finely hairy beneath. Flowers in small clusters. Sepals 5, nearly separate. Petals white, tubular below, the upper part with 1 lobe above and a 3-lobed lip below. Fruit a flat, asymmetric, somewhat fleshy, 1-seeded pod that is finely hairy. Rare. Sandy and silty shores in salt and brackish water; sTX; eLA–MS; FL. Most of year.

LAMIACEAE
Mint Family

Conradina
Conradina canescens (T. & G.) Gray

[574]
Shrub to 50 cm tall. Leaves opposite but with axillary clusters of leaves masking this arrangement without close inspection. Blades narrow, the margins tightly rolled inward so that the blades appear linear or nearly so. Sepals united, hairy, the lobes irregular. Petals united, the upper lip hood-like and much smaller than the 3-lobed lower lip. Corolla white with light-maroon to maroon dots on the lower lip. Fruit 4 nutlets inside the persistent tubular calyx. Common. Dry places—dunes, thin live oak woods, scrub, sand pine, sandy flats, pine-palmetto; AL–wFL. Sept.–Apr.

SOLANACEAE
Nightshade Family

Christmas-berry
Lycium carolinianum Walt.

[575]
Sparingly branched shrub to 3 m tall. Short branches often rigid and thorn-tipped. Leaves simple, succulent, alternate, often with small clusters of leaves from axils, to 3 cm long but usually under 2 cm, linear-oblanceolate to club-shaped, at least a few usually persisting through winter. Flowers solitary from leaf axils. Corolla white to blue or light reddish purple, base tubular, lobes slightly longer than tube, the throat hairy, Fruit a shiny red ellipsoid berry 8–15 mm long, abundantly eaten by birds. Occasional. Sandy mounds and ridges in salt marshes, brackish shores, brackish ditches; TX–sGA. Most of year.

BIGNONIACEAE
Trumpet-creeper Family

Cross-vine
Bignonia capreolata L.

[576]
Vine climbing by leaf tendrils bearing adhesive discs. Leaves opposite, semi-evergreen, composed of 2 entire ovate to oblong leathery leaflets and a branched tendril. Flowers in axillary clusters of 2–5. Calyx bell–shaped, the margin without lobes but with 5 small points. Corolla tubular, irregular, orange to red outside, yellow to red inside, 4–5 cm long. Fruit a flattened capsule to 15 cm long. Seeds flat, many, winged. Occasional. Thickets, thin woods, fencerows; eLA–sVA. Mar.–May. Syn.: *Anisostichus capreolata* (L.) Bureau.

Trumpet-creeper; Cow-itch
Campsis radicans (L.) Seem. ex Bureau

[577]
Vine trailing or climbing by aerial roots. Leaves opposite, deciduous, with 7–13 leaflets. Leaflet margins coarsely serrate. Flowers in terminal clusters. Calyx narrowly bell-shaped, with 5 acute lobes about 5 mm long. Corolla tubular, irregular, orange to red outside, yellow to red inside, 6–8 cm long. Fruit a long narrow capsule tapering at both ends and usually curved. Seeds many, broadly winged. Common. Dry to wet habitats—woods, fencerows, thickets, old fields, stable dunes; eTX–NJ. May–Oct.

RUBIACEAE
Madder Family

Buttonbush
Cephalanthus occidentalis L.

[578]
Deciduous shrub or tree to 8 m tall with leaves opposite or in whorls of 3 or 4. Blades entire, glabrous above,

sparsely to densely hairy below. Stipules short-deltoid, on stem between petiole bases, soon falling but leaving a line as a scar. Flowers many, in dense globose heads about 3 cm across. Corolla white, tubular with 4 short, rounded, spreading lobes. Stamens separate. Ovary inferor. Fruits are angular nutlets densely compacted into globose heads. Common. Moist to wet habitats—pond margins, ditches, swamps, freshwater marshes, sloughs; TX–NS. June–Aug.

CAPRIFOLIACEAE
Honeysuckle Family

Common Elderberry
Sambucus canadensis L.

[579]
Deciduous shrub or rarely a small tree to 6 m tall. Year-old twigs with large lenticels. Pith large, half or more the diameter of year-old twigs. Leaves opposite, compound, with 5–11 leaflets, the 2 lower divisions sometimes composed of 3 leaflets each. Leaflets sharply and rather finely serrate. Inflorescence a large flat-topped cyme to 25 cm broad. Flowers fragrant, 3–5 mm wide. Fruit a purplish to black, juicy, berrylike drupe 4–6 mm long, with 3–5 small one-seeded stones. Common. Swamps, ditches, sloughs, pond margins, depressions; TX–NS. May–July.

Plants with the twice-compound leaves are considered by some to be a separate species, *S. simpsonii* Rehd., or a separate variety of *S. canadensis*.

Arrow-wood
Viburnum recognitum Fern.

[580]
Seaside *Viburnum* species are deciduous shrubs or small trees with opposite leaves. Stipules and their scars are lacking. Bundle scars 3. Petals 5, united, deeply 5-lobed, the lobes spreading. Ovary inferior. Fruit a drupe with 1 small pit.

This species has glabrous twigs.

Lateral veins of leaf blades are straight or nearly so and end at tips of the marginal teeth. Blades with 4–22 prominent teeth on each side. Drupe 5–10 mm long, dark purple to black, subglobose to ovoid, pit globose-ovoid with a broad shallow groove. Occasional. Wet thickets, borders of woods, swales; MD–NBr. May–July.

V. dentatum L. is much alike but twigs are hairy and the drupe pits are ellipsoid-ovoid with a deep groove. Occasional. Similar places; VA–sMA. June–Aug.

Wild-raisin
Viburnum cassinoides L.

[581]
Deciduous shrub or small tree to 5 m tall. Winter buds with 1 pair of outer scales. Leaf blades entire to finely toothed, without marginal hairs, with lateral veins branching and joining before reaching the margin. Peduncles 5–16 mm long. Mature fruits purple to black, 6–9 mm long. Occasional. Swamps, thickets, swales; NY–Nfld. June–Aug.

V. nudum L. (Possum-haw) is quite similar but leaf blades entire to slightly undulating, the margin with at least a few very small hairs. Peduncles 15–45 mm long. Occasional. Similar places; wFL; SC–NC; DE–NY. Apr.–May.

Japanese Honeysuckle
Lonicera japonica Thunb.

[582]
Twining vine with simple opposite deciduous or semi-evergreen leaves. All leaves separate from each other. Blades entire or occasionally lobed. Stipules absent. Flowers and fruits 2 (or by abortion 1) on peduncles in leaf axils. Corolla 3–4 cm long, white or rarely pink, turning yellowish with age. Ovary inferior. Flowers very fragrant. Fruit a juicy berrylike drupe 4–6 mm long, black, with 3–5 one-seeded stones. Common. In almost any habitat except active dunes; TX–

sMA. Apr.–June, sporadically into Sept.

Coral Honeysuckle
Lonicera sempervirens L.

[583]
Twining vine with simple opposite deciduous leaves. A few leaves may persist through mild winters. Uppermost leaves, especially those below the flowers and fruits, united around the stem and thus perfoliate, an unusual feature. Flowers and fruits in terminal clusters. Corolla 35–55 mm long, the tube slenderly trumpet-shaped, the lobes nearly equal. Ovary inferior. Fruits red, juicy, about 7 mm across. Occasional. Thin woods, climbing shrubs and fences, sometimes trailing; eTX–MD. Mar.–July.

ASTERACEAE
Composite Family

Silverling; Groundsel-tree
Baccharis halimifolia L.

[584]
Much-branched shrub or tree to 5 m tall. Twigs long remaining green.

Current year's twigs about 8-ribbed. Larger leaf blades with a few to several conspicuous teeth. Leaves alternate, tardily deciduous, grayish green, glandular-punctate. Stipules absent. Flowers in heads surrounded by bracts, unisexual, sexes on separate plants. Ovaries inferior. Fruits, as in all *Baccharis* species, are cylindrical, or nearly so, achenes tipped with straight bristles. Collectively, the bristles give a female plant a satiny white appearance in autumn. (See photo of *B. angustifolia*.) Male plants can be distinguished from a distance when in flower by a dull yellow cast as in the photograph of *B. halimifolia*. Common. Freshwater marshes, margins of salt marshes, swales, pond shores, fencerows, old fields; eTX–MA. Aug.–Oct.

B. glomeruliflora Pers. is similar in many respects but the leaves are scarcely or not glandular-punctate and the flowers are scattered along the leafy branches instead of being at the ends of branches as in *B. halimifolia*. Rare. Poorly drained to wet places in thin woods or in the open; FL–sNC. Sept.–Nov.

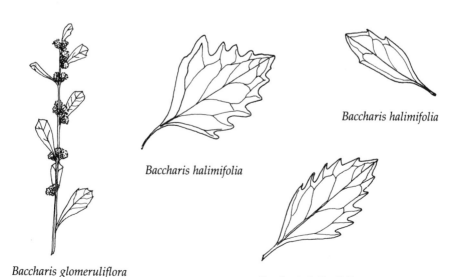

Baccharis halimifolia

Baccharis halimifolia

Baccharis glomeruliflora

Baccharis halimifolia

False-willow
Baccharis angustifolia Michx.

[585]
Much-branched shrub to 4 m tall.
Much like *B. halimifolia* except leaves
are shiny green, nearly linear to narrowly elliptic, 1–5 mm wide. Common. In edge of salt and brackish
marshes, brackish sloughs and
swales, overwash areas, spits; MS–
NC. Sept.–Oct.

Marsh-elder
Iva frutescens L.

[586]
Bushy-branched shrub to 3 m tall.
Soft-wooded, the extremities of the
plant often dying during winter.
Leaves opposite, those on flowering
branches sometimes alternate. Blades
lanceolate, minutely glandular, with 3
main ascending veins, the surfaces
appressed-hairy. Flowers in heads
surrounded by bracts, the functional
stamens and pistils in separate flowers but in the same heads. Ovary inferior. Fruit an obovoid achene 2–2.5
mm long, resin-dotted, lacking
bristles. Common. Brackish or salt-
water habitats—marsh margins, mud
flats, ditches, sloughs, swales; eTX–
NS. June–Oct.

Seashore-elder
Iva imbricata Walt.

[587]
Bushy-branched glabrous shrub to 1.2
m tall with extremities of the plant
often dying during winter. Lowermost leaves opposite, the midstem
and upper ones alternate. Leaf blades
succulent, oblanceolate to elliptic, entire to serrate or dentate, indistinctly
3-nerved from the base. Flowers in
heads 6–8 mm long, surrounded by
bracts, the functional stamens and
pistils in separate flowers but in the
same head. Ovary inferior. Fruit an
achene 3.5–5 mm long, resin-dotted,
lacking bristles. Common. Sandy
beaches, dunelets, dunes, overwash
areas; eTX–VA. Aug.–Oct.

Sea Ox-eye
Borrichia frutescens (L.) DC.

[588]
Little-branched shrub to 1.6 m tall,
mostly around half as tall, spreading
by rhizomes as well as seeds. Young
stems covered with dense gray hairs.
Leaves opposite, thick, somewhat
fleshy, obovate to oblanceolate or spatulate, entire to sharply toothed.
Flowers many, in heads surrounded
by bracts, ray flowers 12–30, the rays
yellow. Receptacular bracts hard and
rigid, with hard sharp erect spine tips
1–3 mm long. The sharp spines and
very compact heads make it difficult
to break open the heads. Fruit an
achene 3–4 mm long, 3–4 angled,
metallic-gray. Pappus segments
united forming a cartilaginous
toothed crown. Common. In margins
and other upper portions of salt and
brackish marshes; TX–VA. Nearly all
year.

Index

Numbers refer to pages. Page numbers set in bold type indicate illustrations in the color section of the book. / For the common names of plants, decisions to hyphenate compound names were made on the basis of taxonomy. For example, Arrow-grass is not a grass and Butterfly-weed is not classified by the word "weed," so the names are hyphenated. By contrast, because Smooth Alder is an "Alder" and Trailing Bluet a "Bluet," the hyphen in each case is omitted and the names are indexed under the last word.